AF752571

Virtues and Economics

Volume 3

Series Editors
Peter Róna, University of Oxford, UK
László Zsolnai, Corvinus University of Budapest, Hungary

Editorial Advisory Board
Helen Alford, Pontifical University of St. Thomas Aquinas ("Angelicum"), Rome, Italy
Luk Bouckaert, Catholic University of Leuven, Belgium
Luigino Bruni, LUMSA University, Rome and Sophia University Institute, Loppiano
Georges Enderle, University of Notre Dame, USA
Carlos Hoevel, Catholic University of Argentina, Buenos Aires, Argentina
John Loughlin, Blackfriars Hall, University of Oxford and Von Hügel Institute
David W. Miller, Princeton University, USA
Sanjoy Mukherjee, Rajiv Gandhi Indian Institute of Management Shillong, India
Mike Thompson, GoodBrand, London, CEIBS Shanghai, and University of Victoria, Vancouver, Canada
Johan Verstraeten, Catholic University of Leuven, Belgium
Stefano Zamagni, University of Bologna, and Johns Hopkins University - SAIS Europe and Pontifical Academy of Social Sciences, Italy

The series is dedicated to virtue ethics and economics. Its purpose is to relocate economic theory to a domain where the connection between the virtues and economic decisions, as that connection is actually experienced in everyday life, is an organic component of theory rather than some sort of an optionally added ingredient. The goal is to help develop a virtue-based economic theory which connects virtues with the contents of economic activities of individuals, unincorporated and incorporated economic agents. The primary context is Catholic Social Teaching but other faith traditions (especially Judaism, Islam, Hinduism, Buddhism, and Confucianism) will also be explored for their construction of virtues in economic action. Special attention will be made to regulatory and policy issues in promoting economic justice.

The series connects virtue ethics with the core of economic theory and practice. It examines the basic and irreducible intentionality of human activities concerned with the production, distribution and consumption of goods and services. It considers the incommensurability of values as the central problem of economic decision making and examines whether that problem can be overcome by any means other than practical reason.

This series will cover high quality edited volumes and monographs.

More information about this series at http://www.springer.com/series/15627

Peter Róna • László Zsolnai
Editors

Economic Objects and the Objects of Economics

Springer

Editors
Peter Róna
Blackfriars Hall
University of Oxford
St. Giles, Oxford, UK

László Zsolnai
Business Ethics Center
Corvinus University of Budapest
Budapest, Hungary

Includes one non exclusive license agreement - Dave Colander: Is Economics a Moral Science?

ISSN 2520-1794 ISSN 2520-1808 (electronic)
Virtues and Economics
ISBN 978-3-030-06866-0 ISBN 978-3-319-94529-3 (eBook)
https://doi.org/10.1007/978-3-319-94529-3

© Springer International Publishing AG, part of Springer Nature 2018
Softcover re-print of the Hardcover 1st edition 2018
This work is subject to copyright. All rights are reserved by the Publisher, whether the whole or part of the material is concerned, specifically the rights of translation, reprinting, reuse of illustrations, recitation, broadcasting, reproduction on microfilms or in any other physical way, and transmission or information storage and retrieval, electronic adaptation, computer software, or by similar or dissimilar methodology now known or hereafter developed.
The use of general descriptive names, registered names, trademarks, service marks, etc. in this publication does not imply, even in the absence of a specific statement, that such names are exempt from the relevant protective laws and regulations and therefore free for general use.
The publisher, the authors and the editors are safe to assume that the advice and information in this book are believed to be true and accurate at the date of publication. Neither the publisher nor the authors or the editors give a warranty, express or implied, with respect to the material contained herein or for any errors or omissions that may have been made. The publisher remains neutral with regard to jurisdictional claims in published maps and institutional affiliations.

This Springer imprint is published by the registered company Springer Nature Switzerland AG
The registered company address is: Gewerbestrasse 11, 6330 Cham, Switzerland

Preface

The papers collected in this second volume of our *Virtues and Economics* series were delivered at a symposium held at the Las Casas Institute, Blackfriars Hall, Oxford University on 7th and 8th of July, 2017. Papers delivered at previous symposia, also at the Las Casas Institute and the Catholic University of Leuven, were published in our previous volume titled *Economics as a Moral Science.*

These papers concern the moral content of economic life and the capacity of economic theory to accommodate that moral content. Our first volume traced and reflected the transformation of economics as a moral science into an analytic discipline and the consequences of that transformation. The essays in this volume address, from a variety of perspectives, the ontological characteristics of the subject matter of the discipline and challenge the assumption that the methodology of the natural sciences can be productively applied to the economy.

The exclusion of the moral dimension of economic life from the axioms, models, and paradigms of modern economic theory, the notion that the economy is the product of the sorts of law-like regularities that characterise the natural sciences instead of being the product of human will, desire and action have had increasingly troubling consequences. Instead of the sort of stability that the discovery of law-like regularities are expected to secure, the search for analogies between the natural and the social sciences has generated increasing social, economic, environmental and political instability. If the wealth of the world has greatly increased, so has human suffering. A re-examination of the foundations of modern economic theory seems to be needed.

The editors would like to thank the authors of these papers for their contribution and to express their enthusiastic support of the dialogue that has developed among them. We would also like to express our gratitude to the Mallinckrodt Foundation for its generous financial support without which these symposia could not have been organised, and thank the Las Casas Institute for its hospitality and administrative support to the 2017 symposium.

Peter Róna
László Zsolnai

Acknowledgement

The editors would like to acknowledge the financial support provided by the Mallinckrodt Foundation to the workshop held at Blackfriars Hall in 2017.

Contents

About the Contributors

David Colander is Distinguished College Professor at Middlebury College, Vermont, USA. He has authored, co-authored, or edited 30 books and over 100 articles on a wide range of topics. These include *Principles of Economics* (McGraw-Hill), *History of Economic Thought* (with Harry Landreth) (Houghton Mifflin), *Macroeconomics* (with Ed Gamber) (Prentice Hall), *Why Aren't Economists as Important as Garbagemen?* (Sharpe), and *MAP: A Market Anti-Inflation Plan* (with Abba Lerner) (Harcourt Brace Jovanovich). He received his Ph.D. from Columbia University and has taught at Columbia University, Vassar College, the University of Miami, and Princeton University. He has been on the board of numerous economic societies and has been Vice President and President of the Eastern Economic Association and Vice-President of the History of Economic Thought Society. His latest work focuses on economic education, complexity, and the methodology appropriate to applied policy economics.

Bruno S. Frey is Permanent Visiting Professor at the University of Basel. He was Professor of Economics at the University of Konstanz from 1970–1977 and at the University of Zurich from 1977–2012, Distinguished Professor of Behavioral Science at the Warwick Business School at the University of Warwick, UK from 2010–2013 and Senior Professor of Economics at Zeppelin University Friedrichshafen, DE from 2013–2015. He is Research Director of CREMA – Center for Research in Economics, Management and the Arts and Co-Founder of CREW – Center for Research in Economics and Well-Being at the University of Basel. He was Managing Editor, from 1969–2015, and is now Honorary Editor of *Kyklos*. Bruno Frey seeks to extend economics beyond standard neo-classics by including insights from other disciplines, including political science, psychology and sociology.

Daniel M. Hausman is Herbert A. Simon and Hilldale Professor at the Department of Philosophy of University of Wisconsin-Madison, USA. He has taught at the University of Maryland at College Park, Carnegie Mellon University, and he has visited at the Institute for Advanced Studies and the London School of Economics.

Most of his research has focused on methodological, metaphysical, and ethical issues at the boundaries between economics and philosophy, and in collaboration with Michael McPherson, he founded the journal Economics and Philosophy. He is also the editor of "The Philosophy of Economics: An Anthology" (3rd edition 2007). His books include "The Inexact and Separate Science of Economics" (1992), "Economic Analysis and Moral Philosophy" (co-authored with Michael McPherson, 1996), "Economic Analysis, Moral Philosophy, and Public Policy" (co-authored with Michael McPherson, 2006), and "Preference, Value, Choice and Welfare" (2011).

Carlos Hoevel is Professor of History of Economic and Political Ideas, Business Ethics and Philosophy of Economics at the Catholic University of Argentina in Buenos Aires, Argentina. He is also Director of the University's Center of Studies in Economy and Culture and Editor of the Journal of Economic Culture. His publications include "The Economy of Recognition" (Springer, 2013).

Kevin T. Jackson is professor and Daniel Janssen Chair at the Solvay Brussels School of Economics and Management in Brussels, Belgium. He is a scholar in international business ethics, global economic governance, and legal philosophy. His research papers have been published in Business Ethics Quarterly, Journal of Business Ethics, Brooklyn Journal of International Law, Harvard Journal of Law and Public Policy and Law and Philosophy. His *Charting Global Responsibilities: Legal Philosophy and Human Rights* (University Press of America, 1994) was presented as a gift to His Holiness the Dalai Lama by the U.S. State Department.

Arjo Klamer is professor of the Economics of Art and Culture at Erasmus University in Rotterdam, The Netherlands, and holds Chair of Cultural Economics. After acquiring his PhD at Duke University, he taught at several US universities including Wellesley College and George Washington University. He has collaborated with Deirdre McCloskey to promote the rhetorical perspective on economics. His latest book is "Doing the Right Thing: A Value Based Economy" (London, 2017. Ubiquity Press).

Tony Lawson is Professor of Economics and Philosophy in the Faculty of Economics at the University of Cambridge, UK He is a co-editor of the Cambridge Journal of Economics, a former director of the University of Cambridge Centre for Gender Studies, and co-founder of the Cambridge Realist Workshop and the Cambridge Social Ontology Group. Lawson is widely noted for his contributions to heterodox economics and to philosophical issues in social theorizing, most especially to social ontology. His books include "Economics and Reality" (Routledge, 1997), "Reorienting Economics" (Routledge, 2003), and "Essays on The Nature and State of Modern Economics" (Routledge, 2015).

Zoltán Pogátsa is a political economist and Associate Professor at the University of West Hungary, in Sopron, Hungary. His research focuses on the economics of European integration. He has published six books and numerous professional and media articles, and is a regular commentator in Hungarian and international media on issues related to European integration and economic development. His book "Heterodox International Political Economics", offers a number of different perspectives on understanding the global economy of the 21st century.

Stephen Pratten is Reader in Economics and Philosophy at King's College, London, UK. After completing his PhD in the Economics Faculty at the University of Cambridge he became a Research Fellow at the Centre for Business Research before joining King's in 1999. His primary research area is the methodology of economics, focusing particularly on issues related to social ontology and realist theorizing. He is co-editor of the Cambridge Journal of Economics and a Research Associate at the Centre for Business Research.

Peter Róna was born in Hungary and emigrated to the United States in 1956. He obtained his B.A. degree in economic history (cum laude) from the University of Pennsylvania and his law degree from Oxford University (First Class) in 1964. He was an associate of the Washington, D.C, law firm, Arnold & Porter and counsel to the US Department of Commerce before becoming the personal assistant of Lord Richardson, Governor of the Bank of England. He joined the Schroder Group in 1969 as the General Counsel of its operations in the United Sates, and became the President and Chief Executive of the IBJ Schroder Bank & Trust co. in 1985. After the collapse of the Berlin Wall he returned to Hungary, where he managed an investment fund. In 2003 he joined the faculty of Eötvös Lóránd University where he taught public international law. In 2006 he was made an Honorary Professor of the Eötvös Lóránd University. He joined the teaching staff of Blackfriars Hall, University of Oxford in 2009, where he teaches courses in economics and the philosophical foundations of the social sciences. His published articles include a study of the Euro and an examination of the philosophical foundations of economics.

Roger Teichmann is College Lecturer in Philosophy at St. Hilda's College, University of Oxford, UK. He gives tutorials at St Hilda's in Ethics, Knowledge and Reality, Logic and Language, Later Wittgenstein, Early Modern Philosophy, and Frege-Russell-Wittgenstein. His research interests are in Ethics, Mind, Action, Language, Wittgenstein, and Anscombe. His book, "The Philosophy of Elizabeth Anscombe" was published in 2008.

László Zsolnai is Professor and Director of the Business Ethics Center at the Corvinus University of Budapest. He is president of the European SPES Institute in Leuven, Belgium. László Zsolnai's recent books include *The Palgrave Handbook of Spirituality and Business* (2011. Houndmills, UK, New York, NY: Palgrave Macmillan), *Beyond Self: Ethical and Spiritual Dimensions of Economics* (2014. Peter Lang Academic Publishers, Oxford), *The Spiritual Dimension of Business Ethics and Sustainability Management* (2015, Springer), *Post-Materialistic Business: Spiritual Value-Orientation in Renewing Management* (2015, Palgrave) and *Ethical Leadership. Indian and European Spiritual Approaches* (2016, Palgrave-Macmillan). His website: http://laszlo-zsolnai.net

Part I
Introduction

Chapter 1
Ontology and Economics

Peter Róna

"The name and not the thing"
William Shakespeare

"All's Well that Ends Well" Act V. Scene III

Abstract The mind-dependent/mind-independent, the thing/activity as well as the older corporeal/incorporeal reality distinctions are all very helpful in distinguishing the subject matter of the social sciences from that of the natural sciences. The distinctions point to a profound ontological difference between the subject matter of the natural sciences and that of the social sciences, and raises important questions about the applicability of the theory formation techniques, criteria and procedures developed in the natural sciences to social science. Measurability, the foundational criterion of the natural sciences assumes that the 'thing', the object of measurement, is measureable by means that are independent of and do not alter or affect the thing being measured. It also assumes that the unit used to measure can also be used to measure other 'things' that bear a relationship to what is being measured. Without it the subject matter of the inquiry is incommensurable.

What is an economic fact? What is a fact in the system of the axioms, the laws and the models of economic science as a whole? How does experience crystallise into a fact, how is an economic fact experienced and how is that fact condensed into an object of economics? Or is it more likely that fact formation in economics is unlike fact formation in the natural sciences, that economic theory contrives the facts with which theory may be formulated? Put differently, is it more likely, that the objects of economics consist not of some simplified, abstracted or isolated, theory-independent facts, derived from reality, but, rather of a surrogate reality designed to

P. Róna (✉)
Blackfriars Hall, University of Oxford, Oxford, UK
e-mail: peter.rona@bfriars.ox.ac.uk

© Springer International Publishing AG, part of Springer Nature 2018

P. Róna, L. Zsolnai (eds.), *Economic Objects and the Objects of Economics*, Virtues and Economics 3, https://doi.org/10.1007/978-3-319-94529-3_1

generate facts that accord with the tenets of a particlar ideology? If modern economics is about model construction, it seems vitally important to understand how what is thought to be the factual content of such models is obtained.

The mind-dependent/mind-independent, the thing/activity as well as the older corporeal/incorporeal reality distinctions are all very helpful in distinguishing the subject matter of the social sciences from that of the natural sciences. The distinctions point to a profound ontological difference between the subject matter of the natural sciences and that of the social sciences, and raises important questions about the applicability of the theory formation techniques, criteria and procedures developed in the natural sciences to social science. Measurability, the foundational criterion of the natural sciences assumes that the 'thing', the object of measurement, is measureable by means that are independent of and do not alter or affect the thing being measured. It also assumes that the unit used to measure can also be used to measure other 'things' that bear a relationship to what is being measured. Without it the subject matter of the inquiry is incommensurable.

The history of modern economics may be seen as a search for a system that not only renders economic phenomena measurable, and also sufficiently homogeneous so that it is, in its entirety, commensurable. Keynes had no doubt about the futility of this conceit. In analysing the Marshall-Pigou notion of the National Dividend, he observes that it consists of non-homogeneous components that do not reflect unmeasured "changes in *value*…(or) evaluate new equiment against old when, owing to changes in technique, the two are not identical." (Keynes 1936, 39.) Keynes also warns that a solution of the difficulty through yielding to imprecision is a dead-end because it fatally compromises any and all causal explanations.

The early attempts to overcome this problem involved the naturalist efforts of the untilitarians. John Stuart Mill's (1843, 573) claim that "Human beings in society have no properties but those which are derived from, and may be resolved into, the laws of the nature of the individual man" may be said to be the foundational axiom of modern economics, containing both the fundamental claim that the laws of economics are part of the laws of nature and the so far unresolved ways in which economic laws may be "derived from" or "resolved into" the laws of nature. Mill follows Hume to the effect that an 'ought' cannot be derived from an 'is' whose 'law' has been and remains to this day, the basis of the naturalist, value-free modern social sciences even after the passing of the Vienna Circle and the collapse of logical positivism.

The next attempt found its commensuring unit of measurement not in pain and pleasure, but in money. The Millean principle to the effect that human nature prefers a greater gain over a lesser one found its measuring unit in exchange value expressed in terms of money.[1] It correctly observed that corporate decisions, indeed the entire discipline of corporate finance is based on this unit. The move, however, requires

[1]Aristotelean exchange value included that quantity of shoes that could be exchanged for a bed. In modern economic theory barter is excluded, exchange is expressed exclusively in terms of money, so that in the measurement of economic growth, the replacement of barter with money based exchange is recorded as a growth in economic activity.

the transformation of things into values, and the further move that specifies values in terms of *exchange* rather than *use*.[2] This radical and profound move, against which Marx railed so vehemently and eloquently, places economic life beyond the reach of ethical assessment, and creates a realm that, unlike all of the rest of our lives, is deaf and blind to morality. Indeed, the imperviousness of economic theory to morality is seen as compelling evidence of its scientific status.

If the commensurability of things was achieved by replacing their ontological properties with exchange value measured in units of money, the incommensurability of activities was overcome by imputing to them a homogenous unit of labour. Performing a piano sonata and carving up a chicken became commensurable on the basis of the units of labour involved in the activity. How that unit was to be determined was even less clear than how price fluctiations affected the value of things if that value was to be measured entirely by their exchange value, but the goal of depriving human activity of its *telos*, and endowing it with the properties of the inanimate objects of the natural sciences was, by hook or crook, achieved.

But the question remains: if the objects of the natural sciences are *things* found in nature, what are the objects of the social sciences in general and of economics in particular. Neither exchange value nor units of labour is a thing the measurement of which is feasible on a basis with the means available to the natural sciences. The question remains precisely because the moral outcome of this paradigmatic change is so troubling. It is not only a question of the means with which commensurability and measurability is achieved; more importantly, what happened to accounting for the moral content of the action of the economic agent? Why should economic action be granted a priviliged and unique exemption from moral supervision? If the escape of economics from the supervision of moral judgment is well founded, why should any other aspect of life not be exempt from it? Indeed, as the techniques developed by economic theory and its off-shoot, game theory, are applied to ever-growing areas of decision-making, the scope for the virtues and for human judgment diminishes. Therefore, clarity on the ontology of the subject matter of the social sciences seems urgently needed.

How do Mill's 'properties' of human beings compare with the properties of natural objects? Do the objects of nature have properties permitting the formulation of law-like generalisations that the objects of economics lack, or are there properties generally common to both? Or, in the alternative, is it possible to set out a theory of causation that is equally valid for both regardless of the lack of common properties among those objects? Putting the question more generally, does the ontology of economic objects matter when formulating economic laws - indeed, does the ontology of economic objects permit the formulation of laws in the first place - and if so, how so? A quote from Nancy Cartwright's influential collection of essays will illustrate the point. Cartwright writes:

---

[2] For a fine analysis of this move see Meikle, S. „Quality and Quantity in Economics" in Mäki, U. (2001) *The Economic World View,* Cambridge University Press, and Meikle, S. (1995) *Aristotle's Economic Thought,* Oxford Universit Press. See also Arendt, H. (1958 *The Human Condition,* The University of chicago press, pp. 163–5

> Nature is rife with very special causal laws involving these causal relations, laws that we represent most immediately using content-rich causal verbs: the pistons *compress* the air in the carburettor chamber, the sun *attracts* the planets, the loss of skill among long-term unemployed workers *discourages* firms from opening new jobs.. (Italics in the original) (Cartwright 2007, 19–20)

Cartwright raises the question whether "there is one thing - one characteristic feature - that makes a law a causal law" and concludes that there is not, that "there is an untold variety of causal relations" and, further, that they do not have a common element (Cartwright 2007, 19) but, as the quote above shows, they nevertheless all occur in "nature".

The means with which economists set out their explanations of and predictions about the subject they address is the model in which notions such as 'supply and demand' 'equilibrium' 'marginal profit' 'the marginal propensity to consume', or, to quote a recent example, "Unobserved Preference Heterogeneity in Demand Using Generalized Random Coefficients" (7) are explored, where, for example, the authors of the cited article

> prove a new identification theorem showing nonparametric identification of the joint distribution of random coefficients models by not imposing a linear index structure for the regressors... then model unobserved preference heterogeneity in consumer demand with random Barten scales. (Lewbel and Pendekur 2017)

But what sort of reality are these terms and concepts meant to cover? What is the subject matter of the measurement to which 'equilibrium', 'marginal propensity to consume' 'random coefficients' 'regressors' 'preference'etc. are meant to be applied? What sort of properties of the object of inquiry can be represented mathematically? It is, of course, quite possible to construct logical, consistent, mathematically expressed models about anything, but what if the reality of, say, "unobserved preference heterogeneity" is neither logical, nor consistent, nor mathematically representable, but something entirely different, or indeed, what if it altogether does not exist? The collection of essays in this volume explore from a variety of perspectives the subject matter of economics and the methods employed by this discipline.

If the intuition defended in some of the papers in this and the previous volume of our series is well founded, the ambition to subject economic life to the rigours of analytic science – the great project of modern economics - is a misconceived conceit. Its proper subject should not be the search for law-like regularities in its manifestations, but, rather, how we, human being, individually and collectively make decisions about our material well-being. If the economy is manmade, the question surely cannot be what regularities of the natural world can be discerned in it – given that it is not a part of nature – but, rather, how *do* we and how *should* we go about creating it.

This book consists of essays examining in various ways and from various points of view the power and validity of the ontological axiom upon which economic theory is erected. At one end of the spectrum the claim is to the effect, that, *pace* Hume, social science is not about making deductions of any sort from 'what is', but about the generation of the 'what is' from the 'what ought to be', that, instead of examining objects that exist independently of it, social science generates, contrives and constructs such objects from premises that postulate a desired end or purpose. (Peter Róna, Chap. 2) Hume's axiom mistakenly postulates the existence of a 'fact' as if social facts were an object of nature. But the sorts of facts from which an ought cannot be derived are not social facts. Social facts are ontologically different from natural facts (Stephen Pratten in Chap. 3 and Tony Lawson in Chap. 4). The workings of time in the construction of social facts is examined in two studies (Roger Teichmann in Chap. 8 and Kevin Jackson in Chap. 9) At the other end there is the view that mainstream economics is on solid grounds even if its methodology requires certain refinements. (David Collander, Chap. 6) The magisterial summation of a lifetime of study of the subject, and with it, a summary of the issues is set out in Daniel Hausman's paper (Chap. 5). Carlos Hoevel's paper (Chap. 7) explores possible passages to free economics from its positivist trap.

László Zsolnai (Chap. 11) argues that the economic reality is far more complex than mainstream economics presupposes. His suggests that we should minimally consider the following types of economic actors: individuals, organizations, communities, networks, and robots (or algorithms) and among these actors at least four basic types of interactions can be identified: market exchange, bureaucratic coordination, ethical transaction, and aggressive action. Bruno S. Frey's paper (Chap. 10) demonstrates that reactivity is an essential characteristic of economic life, i.e. human beings, acting alone or in groups, systematically respond to economic data, and in particular to interventions by economic policy.

Zoltán Pogátsa (Chap. 12) argues that mainstream economics eliminates questions related to the distribution of resources through the very construction of the abstract model it erects. A further significant flaw of mainstream economics is that it concentrates on the sales markets of already existing goods. By doing so, it eliminates the creation of value. Arjo Klamer's paper (Chap. 13) adds that defining economics as the discipline that studies the realization of values is the first step to perceive economics as a moral science. The next step is the inclusion of shared goods as an economic category. One result is a broader scope of economics, one that allows us to consider the goods and values that are of ultimate importance to people, groups, organizations and societies.

Authors in this volume represent differing – and sometimes conflicting – views on the subject matter of economics. However, the contributions collected here suggest that – contrary to Milton Friedman's (1966, 3–16 and 30–43) famous statement on the irrelevance of the realism of economic assumptions – *ontology does matter* in economics. (Chap. 14) The book stands for a richer and more sophisticated economic ontology which may lead to better decisions and policies serving human flourishing in economic life.

References

Arendt, H. 1958. *The human condition*. Chicago: The University of Chicago Press.

Cartwright, N. 2007. *Hunting causes and using them*. Cambridge: Cambridge University Press.

Friedman, M. 1966. The Methodology of Positive Economics. In *Essays In Positive Economics,* ed. M. Friedman, 3–16, 30–43. Chicago: University of Chicago Press.
Keynes, J.M. 1936. *The General Theory of Employment, Interest and Money*. United States: BN Publishing 2006.
Lewbel, A., and K. Pendekur. 2017. Unobserved preference Hererogeneity in demand using generalized random coefficients. *Journal of Political Economy* 125 (4): 1100–1148.
Meikle, S. 1995. *Aristotle's economic thought.* Oxford: Oxford University Press.
———. 2001. Quality and Quantity in Economics. In *The Economic World View*, ed. U. Mäki. Cambridge: Cambridge University Press.
Mill, J.S. 1843. *A System of Logic.*, 1967. London: Longmans.

Part II
The Importance of Ontology

Chapter 2
Objects of Nature and Objects of Thought

Peter Róna

"The art poetic will be weighed in scales"

Aristophanes The Frogs

Abstract The subject matter of the economy is unlike that of the natural sciences. It is the product of human intentionality, and as such, it inherently reflects the human judgment that brought it into being. The fundamental ontological diffence between objects found in nature and objects resulting from human intentionality are explored. The domestication of the techniques employed in the natural sciences cannot produce law-like generalisations in economics with the predictive and explanatory power that characterises the laws of the natural sciences. Modern economics has been most convincing when it has paid attention to the problems caused by the lack of homogeneity in the reality it examines. Studies addressing small, spatially and temporally well defined issues offering a generous supply of empirically verifiable evidence have been more effective in guiding us as to what should be than those aiming for the formation of law-like generalisations of universal applicability.

The subject matter of the economy is unlike that of the natural sciences because the content of its subject matter consists of incommensurable components that cannot be rendered commensurable without altering their ontological properties. This paper defends the following claims:

(i) The methods and techniques of the natural sciences can be applied only to objects;
(i) Socially constructed objects, unlike natural ones, come into being through the agency of human intentionality;
(ii) The events of economic life are incommensurable;

P. Róna (✉)
Blackfriars Hall, University of Oxford, Oxford, UK
e-mail: peter.rona@bfriars.ox.ac.uk

© Springer International Publishing AG, part of Springer Nature 2018
P. Róna, L. Zsolnai (eds.), *Economic Objects and the Objects of Economics*, Virtues and Economics 3, https://doi.org/10.1007/978-3-319-94529-3_2

(iii) Since economic life consists of purposeful activity, the proper subject of economics is not what is, but, rather, what should be.
(iv) Modern economics is an ideology, not a science.

2.1 Introduction

Relations between things, events and properties are either factual, logical or intentional. Factual relations are logically unstable; logical relations are factually meagre, and intentional relations are of an entirely different nature. The Humean principle, – dividing possible relations between the first two - was elaborated by A. J. Ayer (1963: 209–234, and 1956: 144–165) who branded the third one "meaningless", expresses in a nutshell the predicament of economics in its search for scientific status. Factual relations are those that obtain between facts. Logical relations are those that obtain as the product of a rule of logic. Intentional relations – thought by both Hume or Ayer because they are thought to fall outside the domain of science - are those that are the products of human intention. They pose special problems because value judgments inhere in the explanation of phenomena brought about by human intentionality. But discarding intentional relations will not do; intentionality goes to the very heart of economic action. Materialists, naturalists and logical positivists all thought it possible to construct a theory of social action without it, but failed. It is one thing to urge that researches should check their value judgments at the door, quite another to claim that the object of inquiry itself is free of value judgments when, in fact, such value judgments are intrinsic properties of such objects, and yet a third to claim that explanations or interpretations of intentional relations are possible without a framework containing some value judgment.[1] How can a domain of objects – be they things, events or properties or any combination of them – be delineated in a manner that will yield factual law-like relations, or, if that is not possible, how can objects be crafted so that they may be marshalled into logical relations? What can serve as the foundation of intentional relations? In addition to the issues arising from the Humean dichotomy, there is, therefore, the further complication posed by the presence of intentionality in the domain of the social sciences. This paper is about those complications.

Within the great modernist, post-modernist and post-structuralist project, – the dissolution of the autonomous human subject[2] – economics has performed and continues to perform a salient role. In its drive to be equal to the natural sciences it has adopted naturalism as its guiding method and has turned itself into the social science *par excellence* that has drained human identity of much meaning – indeed, the

[1] As Ayer famously put it, propositions that cannot be verified empirically or by the laws of logic are "neither true nor false but literally senseless". Ayer, A. J. (1946) *Language, Truth and Logic.* Dover Press, New York p. 31.

[2] For a recapitulation and analysis of the modernist project see Archer, M. (2000) *Being Human: The Problem of Agency* Cambridge University Press.

whole point of its mission is to drain it of all meaning - and has substituted for the living self the inanimate model. Whereas the principal assault weapon of the project was the reconceptualisation of language as the dissolver of the self, economics went even further; it eliminated natural language altogether and, instead, posited mathematics – thought to be devoid of all human subjectivity – as the value free language of this purported science as if human agency without the intentionality of the human agent were a possible reality.

From among the many facets and ramifications of this project this paper offers an analysis of the 'epistemic fallacy', to wit, the collapse of ontology into epistemology where (i) the subjective experience of the human being is reified, turned into a quantifiable abstract object, (ii) the being of that subjective experience is not only subordinated, but is outright displaced by the 'known' quantified information ascribed to the object, and (iii) the human subject is coerced into a life the contents of which are specified, in the name of rationality, by the object contrived in step (ii). These issues are examined through an analysis of the claim that economics is a science and the associated claim that any knowledge is valid only if it is 'objective' and 'value-free'. (These two requirements, as we shall shortly see, are not co-terminus.)

The final section of the paper addresses the function and purpose of social science. Whether one conceives the social sciences as an analytic, a descriptive or an interpretative discipline, whether it is seen through a naturalistic or an anti-naturalistic perspective, it addresses what *is*. But is there an 'is' which is separable from intentionality, or is social life inextricably as much about what ought to be as it is about what is?

'Objectivity' as the etymology of the word implies, is some sort of knowledge about an object. Objectivity cannot be had about the inexistent nor about experiences that do not involve an object. This paper defends the claim that, for the most part, economics does not have objects that exist outside of its theories, that is to say, its objects are objects of thought rather than objects of nature. But objectivity cannot be had about an object that is the very creation of that purported objectivity. Objectivity and value neutrality are achieved not by some sort of a neutral stance between and among incommensurable values and phenomena, such as, for example, that of Adam Smith's impartial observer, but, rather, by the elimination of incommensurability. Value neutrality is achieved through the extirpation of values.

Modern economics, in compliance with the requirements of modern science, is said to have both explanatory and predictive powers about the behavior of economic agents, the performance of particular segments of the economy as well as of the totality of all of its components. The claim of predictive and explanatory powers rests upon the conceit that the methods and the concepts of the social sciences can be of the same kind as those of the natural sciences because they can be reduced to the relevant theories and methods of the natural sciences, in short, upon the alleged unity of the social and the natural sciences, and that, accordingly, mathematics is capable of detecting and expressing law-like regularities for both.

The claim advanced in this paper is that the social sciences in general and economics in particular is unlike any of the natural sciences because:

(i) Science and its methods can only be applied to objects;
(ii) Socially constructed objects are ontologically different from those found in nature because they come into being through the agency of human intentionality;
(iii) the subject matter of the social sciences in general and of economics in particular – the actual human experience in the making and consumption of goods and services - does not consist of or contain objects that resemble the objects of nature. Whereas the objects of nature are known by their empirically verifiable, measurable, properties economic objects, such as work and shopping are events and activities. The objects of economics, said to summaries and represent events and activities, such as 'demand', 'equilibrium', 'the marginal propensity to consume', etc. are abstract entities. They are objects of thought;
(iv) The singular events of economic life, unlike the properties of the objects of nature studied by the natural sciences, cannot be generalized without, in effect, substituting for the singular event or group of events an abstract entity. They are incommensurable. Whereas the valence of an element in the periodic table holds true for all such elements, – and because it holds true for each instantiation of that element it holds as well for all instantiation of a valence value – the actual purchase by Jones, or a group of people of a pint of milk into a generalized demand curve for milk involves the replacement of a concreta with an abstraction;
(v) Whereas social science methodology is said to be concerned with how, in the famous phrase of J. S. Mill through the processes of idealization, simplification and reduction, the 'disturbing causes' are removed from social phenomena so as to reveal their defining causal mechanisms, the claim here is that, in fact, it is not these processes that characterize modern economics, but the outright substitution of an alternative or surrogate reality in place of empirical evidence. Supply and demand curves do not exist, and there is nothing in ordinary experience from which, with the removal of disturbing causes, they would become manifest. This substitution takes place because the abstract entities with which theory is formed – due to their fundamentally different ontological characteristics – cannot be derived from the evidence the theory purports to explain. These contrived abstract entities, like abstract entities in general, have no intrinsic causal powers, and, unlike the empirical evidence they purport to summarize and represent, lack spatial and temporal specificity. Causal powers are either ascribed to them by stipulation or by the operation of mathematics on the untested assumption that mathematical relationships validly represent the relationships that arise or exist between economic objects.
(vi) Economic phenomena are conglomerates of incommensurables that cannot be forced into an algorithm. Modern economic theory produces compelling idealized results from mathematically representable commensurable - meaning sharing a common unit of measurement – components, but cannot cope with the inherent incommensurability of economic phenomena.

Because the question of commensurability lurks in the background and infuses all of these issues, because a showing of incommensurability produces entirely different results from its opposite, and because modern economic theory is built on the contrivance of perfect commensurability, the first section of the paper deals with this age-old problem.

2.2 Value Neutrality and Incommensurability

The connection between counting, weighing and measuring on the one hand and comprehending on the other was well recognized by the time of Plato. (Nussbaum 2001: 107–110) Indeed, the Pythagoreans, much like Lord Kelvin and 20th century positivists, aggressively maintained that knowledge could be had only of numerically expressible objects even if there was wide divergence of opinion among them about what objects were, in fact, numerically expressible.[3] What is quantifiable and what is not, or, to put it differently, what is commensurable and what is not is the foundational question of economics. If the hallmark of a science is a subject matter that can be measured, economics must concern itself only with measurable objects if it aims to be a science. If it can do so, if economics can identify for its attention an autonomous domain consisting entirely of quantifiable objects and phenomena, its ambition to be an analytic science much like the natural sciences, its hopes for formulating law-like generalizations of universal applicability,[4] rests on promising foundations. But if it cannot, if minimal descriptive adequacy requires it to pay attention to objects that defy quantification, it must either resort to devices with which qualities and values, in the words of Hausman (1992), "are hammered into" quantities, or abandon its ambitions to be seen as an analytic endeavor.

The great predicament of economics arises from the irreconcilable contradiction between the deep human commitment to a plurality of values, entailing as it does disorder and chance, and the desire for order and predictability, in short, Hausman's metaphor. The contradictions and the chaos of the particular given in human experience is resolved in everyday life with the help of practical reason, with the appeal to what should be as opposed to what is. Because practical reason is the world of the unquantifiable, the world of human judgment, desire and will, necessarily

[3] Plato *Republic* 522c "I mean number and calculation, for isn't it true that every craft and science must have a share in that?" Although the ancient Greeks came quite close to devising an instrument with which temperature could be measured, it seems that temperature was not thought to be among the measurables. For a systematic analysis of measurement in ancient Greece see Lloyd, G.E.R. (1987) *The Revolutions of Wisdom* University of California Press, Chapter V and Nussbaum (2001, 107–110).

[4] Lord Robbins famously required only "almost" universal applicability to Karl Popper's great distress. When and where would that „almost" amount to a falsification, or does that „almost" evade altogether the Popperian falsification test? Hausman's version of Lord Robbins' „almost" is reflected in the title of his seminal book. He sees economics as a „separate and inexact" science. (Hausman 1992)

encompassing the world of morality, satisfaction of the imperative to quantify necessarily involves the expulsion of the moral from science's domain. But after the banishment of morality is economics left with a domain sufficiently descriptive of the reality it purports to address? Or, even more frighteningly, when Hausman's hammering is completed, are we confronted with a surrogate reality hitherto unknown to experience, a reality that cannot be comprehended by any extension of human experience but that nevertheless must be obeyed? It is the ability of modern economics to create surrogate realities and then build the legal, political and social institutions with which compliance with the requirements of that surrogate reality are obtained rather than its limited predictive and explanatory power that raises the most troubling questions about the discipline.

The question, undoubtedly a deeply philosophical one, acquires immediate practical significance when an amount of goodwill is set out opposite an amount of liability for borrowed money in the balance sheet of a corporation. The goodwill is quantified, implying that there exists a unit of measurement it has in common with the amount of the liability, that is to say, that these two items are commensurable. In particular, it is claimed that the unit of measurement for the liability also measures goodwill with equal degree of accuracy. But does it? The unit employed to express the amount of assets and the amount of liabilities in fact measure incommensurable realities. As every financial analyst knows, liabilities are real; assets are estimates. The illusion generated by their inclusion in the same model – in this case the balance sheet – is the illusion of commensurability with which all of economic theory, as indeed all of corporate finance is infected.

The leap from the incommensurable and contingent particular to the eternal commensurable is the great theme of Plato's perhaps two most dramatic and spectacular masterpieces, *Protagoras* and *Symposium* with which, among other things, he unwittingly laid the foundations for the modern *homo oeconomicus*. In *Symposium* Diotima persuades Socrates of the value of abandoning the tormenting world of the love of the incommensurable particular, and, instead, to love an abstracted version of the object of love consisting and embodying only those qualities that merit love. Instead of being a part of the particular incommensurable person, the qualities should be removed from its contingent and particular context, and by so doing rendered entirely commensurable, so as to be found not only in many other persons, but even in entirely abstract subjects. Beauty in this conception is a universal of which a beautiful body and a beautiful mathematical proof are its instantiations in which we are to love not the instance but the universal. The love of the beauty of a person is not to differ from the love of the beauty of mathematics.

The construction of commensurability, – Socrates' ascent on the ladder set out for him by Diotema as the means to escape the anguish of the particular - is built on the distinction between *technē* and *tuchē* explored and elaborated in *Protagoras. Technē* is measurable know-how with the help of which the disorderly particulars can be rendered commensurable, and exposure to the risky world of luck and chance, *tuchē,* can be eliminated.

In sharp contrast, the drunken Alcibiades – his drunkenness illustrates the disorderliness of the particular - puts forward the view that human truths are only those that are gained through *gnōnai*, (usually translated 'understanding gained through experience', but better translated here as 'suffering'). The Socratic *epistēmē*, concerned with universals and the deductive method of science, meets Alcibiades' truth about love gained through the particular passion for a particular and contingent loved one. The particular for Socrates, on the other hand, is true if and only if it is the instantiation of a universal truth, and, to make matters even more dramatic, Socrates requires the instantiation to be repeatable in order to qualify as a truth bearing exemplar. The task for Socrates is the construction of a surrogate world where the part of our humanity that grasps reality through *gnōnai* is replaced with *technē* with which measurability is achieved by the elimination of all values except one. Socrates' *technē,* first proposed as the means for overcoming those pesky particularities could be ordered into a commensurable sequence without altering their nature turns out to be the single most powerful tool for the liquidation of our humanity. His *technē* establishes the dictatorship of the single value to which all other values, deprived of their particularity, must and can be related *quantitatively.*

Commensurability, the ordering of facts and values quantitatively requires the maintenance of the fact/value dichotomy because only objects with a unit of measurement in common can be commensurable. But objects that are compounds of fact and value cannot be so measured because the values are not subject to quantitative assessment. What then are the 'facts' of economics? How are they extracted from experience, or, in the alternative, how are they composed or contrived?

For the 18th century a scientific fact was an instantiation of a system, and the system was either of providential design, or, as in Hume, who did not believe in Providence, the consequence of experience. Science then, was systematic knowledge, or more precisely, knowledge of a system, but how could the existence of the system be established beyond it being some sort of an ideal fuelled by the human desire for order, so powerfully shown in Plato's Protagoras? (Poovey 1998) The tension between chance and order or the particular and the general needed for the articulation of an order reappears again and again as confidence in the order producing ability of Providence ebbs. Under the solution developed by Dugald Stewart, the leading Scottish philosopher at the end of the 18th and the first part of the 19th century, fact was obtained through abstraction, defined by him as "This power of considering certain qualities or attributes of an object apart from the rest; or, as I would rather choose to define it, the power which the understanding has of separating the combinations, which are presented to it..." (Stewart 1792). But the definition does not get us very far, because it requires an object for the attentions of this operation. The basic question – how are such objects found or formed – is evaded, throughout the 19th and the twentieth century, and remains insufficiently addressed today.

2.3 Object Formation

How is economic experience turned into economic objects and how are those objects then turned into the objects of economics? Whether economics is a science stands or falls on the sustainability of the claim that its models and its procedures for making models are analogous to the models of the natural scientist, a claim that, in turn depends on whether the subject matter of the model shares the ontological characteristics of the subject matter of the natural sciences. The highly influential work of, among others, Nancy Cartwright (1989) and Mary Morgan (2012) assumes the isomorphism. Other leading theories of the model, such as that of Hamminga and DeMarchi (1994), – where the real world is said to be simplified so as to be suitable for theoretical description - or Gibbard and Varian (1978) – connecting mathematical theory with causal observations about the economy – or Havelmo's seminal paper on the correspondence problems between mathematical theory and statistical data, all assume, with the help of probability theory, the feasibility of the transition from the abstract to the concrete and vice versa. This paper claims the opposite, and proposes instead that the objects in models are surrogates, ontologically different from the reality they displace. Much like allegories, they may and often do illuminate some aspect of the reality for which they are standing in, but these illuminations are short lived parallels. Models are ideologies rather than science, designed to show the consequences of proposed policies formulated in the purportedly scientific language of axioms.

Economic facts are particular experiences obtained in the ordinary course of life in connection with activities undertaken to secure the material means of existence. Their generalization into the sorts of facts of which economic theory is composed cannot be obtained with the employment of the techniques of the natural sciences because they are not the sorts of objects upon which Stewart's or, for that matter, Diotima's procedure can be employed *without first specifying those 'qualities or attributes' of which the selected fact is thought to be an instance.* It is, of course, true that the specification will engender a reflexivity with which the qualities and attributes are marshaled as well as generated, but the specification is the product not of the discovery of some inherent property of the experience, but of human intentionality even if that intentionality takes into account experience.

2.4 Measurability

Measurability as the criterion by which all knowledge was to be validated was adopted by economics with binding force during the second half of the nineteenth century and has remained with us to this day. As Lord Kelvin explained, henceforth *all* knowledge was a matter of measurability:

> I often say that when you can measure what you are talking about, and express it in numbers, you know something about it, but when you cannot express it in numbers, your

knowledge is of a meagre and unsatisfactory kind; it may be the beginning of knowledge, but you have scarcely, in your thoughts, advanced to the state of Science, whatever the matter may be.

But what if, as Michael Polanyi (1958) pointed out 'we know more than we can say'?[5] To make matters even worse, what if we know many things that we can say but cannot measure? What if economic events come about on the back of decisions made with the help of things we can neither measure nor say? If scientifically valid explanatory and predictive power rests on the measurability of the properties of the object of study, what sorts of things can be subjected to measurement on the one hand, and what sorts of things should be but cannot be measurable in order to be the proper objects of science? Do measurable objects consist only of objects found in nature, or can measurement of the sort practiced in the natural sciences also find application to socially constructed facts?

Scientific measurement requires a unit, a method and a means of measurement – typically an instrument – that is wholly independent of both the object being measured and the measurer. It is the result obtained with the help of such instruments that the natural sciences classify as (provisional) facts and lends them the aura of reliability denied to values for lack of such measurability. The absence of an independent measuring instrument should have been seen as a warning sign that something was amiss, that the ambition to catch up with the natural sciences could be defeated by the lack of a measurable object, that, in short, the 'facts' of economics were unlike the facts of the natural sciences. The fact that the valency of a hydrogen atom is +1 is not like the fact that the marginal utility of the second consumed unit is less than that of the first. But it was thought that the mathematisation of the subject matter – that is to say the deployment of numbers, charts and graphs in models as the subject matter of study – could overcome the problem, inasmuch as what was to be studied were not objects with measurable properties but the purported measurements themselves. The requirement that the measurements and any causal relations they disclosed would still have to be related back to the empirically identified object was and remains to be quietly ignored; we end up with a demand curve without identifiable demanders, a general equilibrium balancing between and among unknown and unknowable components, etc.; the bridge between the concrete and the abstract remain even if, as we will see, an intermediate abstraction has been inserted.

It will be seen that much of economics rides on a notion of 'fact' inherited from the logical positivists and largely unexamined in economics,[6] even though the fact/

[5] In asking this question Polányi follows Adam Smith and Karl Marx, who were serious Hellenists, the first a Stoic and the second an Aristotelian for whom the sharp Platonic distinction between technē and tuchē, described above, was unconvincing.

[6] The work of Vivian Walsh, Hilary Putnam, Amartya Sen and Martha C. Nussbaum is a major exception this generalisation, but their work has been largely ignored by mainstream economics. See Putnam, H and Walsh, V.: *The End of Value-free Economics*. 2012. New York and London, Routledge.

value distinction was demolished by Quine (1951). As Hilary Putnam (2002: 9) writes:

> A version of … the fact/value dichotomy ("is" versus "ought") and the analytic/synthetic dichotomy ("matters of fact" and "relation of ideas"), was foundational for classical empiricism as well as for its twntieth-century daughter, logical positivism. Thus to come to think without these dogmas is to enter upon a *genuine* "post-modernism – to enter into a whole new field of intellectual possibilities in every important area of culture. (Emphasis in the original.)

Instead of challenging this reductionist and determinist epistemology as unsuited to the material to be studied, modern economics has built its domain of knowledge on the assumption that the domestication of the methods of the natural sciences would enable it to contrive for its field of study the sorts of objects that could be the subject of measurement. In the words of Hausman (1992), the imported methodology, it was hoped, could be "hammered into" a mathematically expressible, and therefore measurable form of the matter under study, and provide economics with the sorts of objects that could be deployed in models with the help of which law-like regularities in the behaviour of these objects could be established.

The claim to scientific status assumes that economics has a subject matter with regard to which it is possible to collect in an objective manner verifiable data, and that the data so collected is *homogenous with the reality under study, that it not only represents in some sense that reality, but that it is part and parcel, a genuine instantiation of the reality itself rather than some sort of a concept obtained through some sort of a distillation of the purported essence of that instantiation.* When Robin Findley Henry defines chemistry as the "attempt to understand transformations between substances", the subject of his study is a "substance", a corporeal thing that remains constant, and "transformation" is a collective term for the recordable changes that occur when substances come into the sort of proximity that triggers them.

When Lionel Robbins defines economics as "the (almost) universally applicable science of choice in conditions of scarcity" the subject matter of his purported universal science, "choice" and "scarcity" are concepts without a stable reality behind them. It is, of course, quite possible to collect data about events taking place at a particular place at a particular time. But how can such data be generalised into the sort of law-like regularities that are the essential feature of the hypotheses of the natural sciences? How can the evanescent and incorporeal reality of economic life be reified so that its subject matter resembles the chemist's substances?

The attempted answer to this question is the foundational tool of modern economics: model making. Models stabilise the contingent reality of economic life by authorising the crafting of stable objects with the help of (i) axiomatic assumptions, including the stipulation of causal relationships, (ii) isolation, (iii) idealisation, and (iv) abstraction. The claim to scientific status therefore stands or falls on the truth value of economic models, and that truth value stands or falls on the empirical validity of these procedures. Although Milton Friedman (1953: 8-9) famously claimed the opposite, namely, that "the only relevant test of the *validity* of a hypothesis is comparison of its predictions with experience", – a requirement that, in any event, no economic model has met so far – successful fortune tellers and hedge fund managers are still not considered scientists, and with good reason. A scientific prediction must rest on a law-like regularity, and the law-like regularity must have empirical validation not only with respect to its predictions but also with regard to the mode, the path leading to the fulfilment of the predictions. A prediction that a piece of paper will catch fire when thrown onto a burning pile is not a scientific hypothesis. Authorities responsible for the licensing of medicinal compounds do not, as a general rule, issue such licensing in the absence of an understanding of the mode of action of the compound. Exempting economics from this second requirement means that economics does not share the scientific method of the natural sciences, but, rather, only mimics them in its search for credibility.

2.5 Measuring

Model making requires quantification, and quantification requires measuring. It is generally agreed, that some phenomena may be measured experimentally, while others, those that can be based only upon observation, should be measured statistically. At the same time, the possibility, that economic life might consist of incommensurable components, or, *horribile dictu* even some elements that cannot be reduced to quantification at all, that cannot be measured, is anathema to economic theory. The admission of incommensurability or unmeasurability would undercut the scientific status of economics. Measuring and the justification of measuring techniques therefore occupies a central place in economic theory. How do economists measure the phenomena they study in comparison with the natural scientists? Does the availability of a physical object to the natural sciences and its unavailability to the economist with which measuring is accomplished matter? What is the connection between reality and its purported numerical representation?

Some students of the subject, such as Lucien March, felt that the difference between the experimental sciences, where one element that exists among several in a phenomenon can be isolated, and the observational sciences where a fact, having occurred in historical time and space, can never be replicated, and the context,- the so called "adjacent circumstances"- cannot be fully identified let alone recreated, means that the method suitable to the one is inappropriate to the other.[7] March saw two major obstacles to the crafting of common measuring techniques for the experimental and the observational sciences. First, the sheer complexity and the indeterminable time parameters of the facts at play made the isolation of one or more facts for study futile. Second, the hope of constructing any sort of a causal theory for the observational sciences seemed equally futile. Instead, he thought that the careful study, definition and collection of facts permitted the identification of correlations,

[7] For a fine analysis of March's *oeuvre* see Jovanovic, F. And Le Gall, P. „March to Numbers: The Statisctical Style of Lucien March" in Klein, J. L. and Morgan, M. S. (Eds.): *The Age of Economic Measurement*. 2001. Duke University Press.

or, as he called it, "the coefficients of dependency". March essentially followed David Hume and Karl Pearson in holding that in the absence of an internal necessity no absolute regularity can exist between cause and effect. Causation was nothing more than a simplifying image imposed on ever changing Nature by the operation of the mind. History therefore mattered in a way that cannot be accommodated by the techniques of the experimental sciences; correlations, also intrinsically ahistorical, required their combination with graphs,[8] but even such a combination only disclosed fleeting and partial images of reality. Any regularity extracted by the social sciences was evanescent, but not all was lost. In place of the determinism of mathematical economics statistics offered a far lovelier means with which chance, instead of being eliminated, could be subjected to analysis.

It should be noted that for March the issues and problems associated with measuring are epistemological. It is the sheer complexity of the phenomena that precludes the sort of measurement that is available to the natural scientist that makes the task so difficult because, among other things, complexity frustrates the isolation of causes, and it is the ability of statistics to deal with complexity on a probabilistic basis without confronting the issue of causation that renders it suitable for the study of social phenomena. In short, it is a problem that could be solved with the arrival of Big Data. The idea that the problem is not only epistemological, not merely the consequence of complexity, but, rather, stems from the radically different ontological characteristics of the natural sciences on the one hand and the social sciences on the other does not occur to him, and, as we shall see, escapes the attention of all of the leading students of the subject – from Jevons, Walras, or Fisher to our contemporaries, including such eminent scholars as M. S. Morgan, H. Maas, J. L. Klein or M. Boumans.

Since the days of Lucien March the view that there are no meaningful differences between the "measuring instruments" of the natural and the social sciences has regained the dominance it enjoyed from the introduction of the metric system – which, it was hoped, would render commensurable any and all phenomena in the physical and the social world alike – until that brief period between the First and the Second World Wars and perhaps the decade following it when the incommensurability of phenomena occupied the minds of Keynes and his followers. Thereafter, the ontological concerns of Keynes and the Cambridge Circus – particularly Kaldor (1974) and Robinson[9] (1981) – were largely abandoned under the pressures of positivist philosophy which, for the remainder the twentieth century rendered ontology disreputable.

Although the high hopes for the metric system soon met its skeptics – J. S. Mill, for example, had no confidence in the possibilities of measurement and mathematisation despite his commitment to the applicability of Newtonian physics to economics – W. Stanley Jevons (1911) never waivered in his belief that what is

[8] An examination of the question whether March's claim that the combination of correlations with graphs salvages spacial and temporal information is beyond the scope of this paper.

[9] See for example J. Robinson: *What Are the Questions? And Other Essays.* (1981.) for the growing concerns of Cambridge UK in its opposition to Cambridge Massachusetts. M.E. Sharpe Inc. New York and the references therein.

sauce for the goose is sauce for the gander.[10] From the earliest years of his career he was fascinated with all types of measuring instruments, including the barometer, the hydrometer, the thermometer, and especially the balance.[11] He built a number of them, and considered that it was the measuring instrument that gave birth to a science rather than the other way around. He found in the balance the physical tool and in what that tool measures, namely equilibrium, the central organizing instrument of his thoughts without realizing that he was converting an actual instrument into a metaphor. He converted Bentham's felicific calculus, where pain and pleasure were motivations, into physical forces, measurable with the help of the balance, and comparable when expressed numerically in the same way as the weights placed on the two trays of the scale. He illustrated his theory with precise drawings and showed how the problem of multiple causes could be avoided by placing all of them on the same tray, much as a loaf of bread and a bucket of coal could be measured together by the weight placed in the other tray. Jevons' confidence in the commensurability of physical and social phenomena reached its apex in his studies of the causal connection between sun spots and the business cycle.

The idea that measuring instruments could be contrived to measure economic phenomena in much the same way as physical instruments are built to measure physical phenomena reappeared with increasing vigor after the Second World War, much supported both ideologically and financially by the exigencies of the Cold War.[12] A showing of the scientific, and therefore objective nature of the capitalist system as opposed to the subjective ideology of socialism was seen as conclusive proof of the superiority of the former. The purported objectivity of the purported tools with which economic phenomena was to be measured has performed and continues to perform a critical role in the legitimization of capitalism.

What are the "measuring tools" of economics and how are they conflated with the measuring tools of the natural sciences?

2.6 Model Making

The following is a partial investigation of the scientific foundations of model making concentrating how the objects deployed in the model are located or come into being. The model making techniques of economics, all taken from the natural sciences, are isolation, idealization and abstraction.

---

[10] Mill, J.S. (1843) *System of Logic*. Longmans, Green & Co. London, Book III, Ch. VI. Sec. "Introduction" and Jevons, W. S. *The Theory of Political Economy*. Palgrave Macmillan pp. xxxviii–xxxix.

[11] For a detailed analysis of Jevons' interest in measuring instruments see Maas, H. „Jevons' Balancing Acts" in Klein, J. L. and Morgan, M. S. (2001) *The Age of Economic Measurement*. Duke University Press

[12] (13) The U:S: Department of the Navy was an important source of financing for this research.

2.6.1 Isolation

Isolation, perhaps the most common procedure in the natural sciences, is the procedure whereby from among a group of interrelated elements one is selected for the study of its properties, and experimentation is undertaken where the separation is to be accomplished without causing any change to the selected element,[13] and the removed elements are identified and accounted for. J. S. Mill's "disturbing cause" is removed so as to permit the analysis of what would have been disturbed had that pesky cause been permitted to remain. It is in later and fuller form the distinction between "endogenous" and "exogenous" elements. But elements or phenomena do not come with any of these labels. How does one decide what is a disturbing cause in the absence of physical separability, or an exogenous element without a theory that posits this distinction?

In physics the answer to this question is rather simple: the element is removed because (i) it is not the subject of investigation, and (ii) proof that its removal does not affect the investigation of the chosen subject is supplied by accounting for the consequences upon its reintroduction. But isolation in, say, physics and in economics is not the same, and the differences can be shown by a comparison of Newton's law of inertia with the economist's law of rational choice. Newton's first law refers to what happens to a body that is not subject to any external force. Rational choice theory assumes a rational agent. Are these two assumptions justified by similar means? Put differently, is the truth value of Newton's first law independent of the truth value of the assumption of no external forces, whereas the truth value of the theory of rational choice dependent on the assumption of rational agents? The absence of external forces is an assumption in the sense that in the observable world all bodies are subject to external forces, but it is not an assumption in the sense that it is quite possible to imagine and now to create the absence of external forces. The rational agent is not an assumption based either on the removal of "irrational" elements in his behavior so as to be left with what remains or on the idealization of empirical evidence; his "rationality" is defined entirely by the requirement of the model. It is not the independently identifiable rationality of the agent that determines the truth value of the model, but, rather, the requirements of the model that define the rationality of the agent. The state of inertia enables us to measure the workings of external forces that, but for its isolation, could not be measured. Inertia is a concrete state that first could be imagined based on empirical evidence, and now actually replicated. But "rationality" is not a concretum, but an abstraction where what is obtained is not what has been isolated after the removal of empirically obtainable elements, but, rather, the invention of something that does not in fact exist.[14] It cannot be falsified because no empirical evidence is relevant to it. Instead

[13] An experiment may, of course, address the phenomena attendant upon separation, but an analysis of these cases is not relevant to economics.

[14] A robot could not be made into a rational agent without specifying the terms of its rationality and confining his choices to a limited set.

of meeting the requirement for empirical validation of the objects so contrived, they are validated simply by the operation of the model.

The Millian claim, advanced by Musgrave (1991: 377–387) in an effort to save Friedman's argument justifying unrealistic assumptions, restated by Cartwright (1989) and more recently by Mäki (2002: 317–335) on the basis that disturbing causes can be identified and removed so as leave for investigation causal factors of interest – recreating for economics circumstances analogous to laboratory conditions - reasons on the basis of the ontological characteristics of objects of nature combined with patently arbitrary assumptions. They all assume, one way or another, that the procedures for isolating components or elements of natural objects will operate similarly in the case of incorporeal objects. But incorporeal objects, as Adam Smith knew, cannot be decomposed; the attempt to decompose them changes their ontological properties, and reliance on unreasonable assumptions is no more justifiable in economics than in mathematics. As Bertrand Russell noted (1919) "The method of postulating what we want has many advantages; they are the same as the advantages of theft over honest toil."

The objects of thought, customarily referred to as incorporeal objects in the eighteenth century, are unlike the objects of nature, or corporeal objects. In short, compliance with the nineteenth century conceit about the unity of the natural and social sciences is achieved by crafting a surrogate reality that mimics the ontology of the objects of the natural sciences and leaves behind social reality.[15]

The ambitious hope of the Enlightenment, that a common language, to wit, the language of mathematics binds the natural and the social sciences together, that despite the ontological differences, mathematics could represent both and supply the logic with which either could be analyzed and understood could not escape unscathed from the 19th to the twentieth century. The claim of mathematics to truth in the sense of inherently representing the laws with which physical reality is governed turned out to be no more than first a lovely dream and then the nightmare of totalitarianism (Arendt 1951). The threefold tension between and among logical truth, factual truth and numerical representation – already foreshadowed in Leibnitz's distinction between necessary and contingent truth – broke into the open with the birth of non-Euclidian geometry.[16] It is a curious fact that economics entrusted its fate to mathematical modeling just as the discipline's representational capacity came under serious questioning, namely, well after the publication of Hermann Helmholtz's (1887) *Counting and Measuring* in which the author rather conclusively refutes the notion that some inherent property of arithmetic renders it capable of automatically, that is to say, without independent experience, reflect physical phenomena. Helmholtz showed, that there is no *a priori* way to insure that arithmetic applies to any given event, substance or circumstance. The fit between

[15] This need to craft a surrogate reality in order to model incorporeal reality also spells trouble for materialism, but that subject is beyond the scope of this paper.

[16] For an excellent recounting of this history see Kline, M. (1980) *Mathematics; the Loss of Certainty,* Oxford University Press.

mathematics and the objects to which it is applied has been and remains one of the oldest puzzles. As Albert Einstein noted:

> Here arises a puzzle that has disturbed scientists of all periods. How is it possible that mathematics, a product of human thought that is independent of experience, fits so excellently the objects of physical reality? Can human reason without experience discover by pure thinking properties of real things? [17]

The eminently reasonable answer to this quandry has been the acknowledgement that we do not know, and precisely because we do not know, we need to perform the experiments with which the results obtained with the aid of mathematics can be confirmed. The need to do so remains whether one takes the Kantian view that mathematics is not something independent of the external world, but, rather, our means for conceiving it, that it is the only means with which we can impose intelligible order on chaotic sense perception, or the more traditional one to the effect that nature is constructed mathematically and mathematics is nothing more than the decoding of laws enbedded in nature. In either case the need for the connection between nature and mathematics remains frustratingly probabilistic; empirical evidence is the meagre substitute for the missing theoretical proof.

Even in the most basic case, the design of the *homo oeconomicus*, what at first sight appears to be a case of straightforward simplification – namely the elimination of certain identifiable features so as to remain with the one to be formed into an object –, turns out on closer examination to be an invention. Adam Smith (1776) depicted the economic behaviour of his protagonist as the product of a complex mixture of the naturally given propensity to truck, barter and exchange, self-interest, talents, sentiments and tastes *combining in unstable and ever-changing proportions.* By the time we get to John Stuart Mill, economic man's behaviour is entirely accounted for by his preference of a greater gain over a lesser one. It may seem that the difference between the two conceptions is that the first one is a thick while the second one is a thin description where the latter is obtained by simply eliminating features of the former.[18] The real difference, however, is not in the number of features recognised by Smith and Mill respectively, but the inherent instability of the former resulting from the indeterminate manner and proportion in which those features combine in the formation of economic behaviour as compared to stability obtained when the scope for such combinations is eliminated. The former is not a description that can be turned into an object; it cannot be a component of a model. The latter is contrived to satisfy the stability requirement of the sort of object about which law-like generalities can be formulated; Mill's move produces a fundamental ontological change and of course also heralds the abandonment of descriptive adequacy in favour of analytic clarity.

[17] Einstein, A, (1921) *Sidelights on Relativity,* quoted in Kline, M. (1980) *Mathematics; the Loss of Certainty*, Oxford University Press. Some theories deny that mathematical thought is 'independent of experience'.

[18] This is the generally held view. See Morgan, M. S. (2012) *The World in the Model; How Economists Work and Think,* Cambridge University Press, pp. 138–139.

If the raw material, the subject matter of economics is experience, how does it compare with the subject matter of any of the natural sciences? How can experience or some selected segment of it be made into an object devoid of the subjectivity of experience? As Thomas Nagel (2016: 30) noted:

> If the subjective character of experience is fully comprehensible only from one point of view, then any shift to greater objectivity – that is, less attachment to a specific viewpoint – does not take us nearer to the real nature of the phenomenon: it takes us further away from it.

Economic activity is about the purposeful mental formation of objectives and their pursuit. It is conscious intentionality, a mind dependent activity from its inception all the way through to its realisation that makes sense only from the point of view of its protagonist. Without the subjective intentionality of its protagonist it does not come into being. The objects of nature, on the other hand, exist quite independently of human intentionality. They are mind-independent unaffected by human perception. The success or failure of economics as a science depends on its ability to transform these mind-dependent or intentional objects of actual experience into objects that are innocent of intentionality. If isolation necessarily involves an ontological category change, if "it does not take us nearer to the real nature of the phenomenon" is there any other way to get from experience to concepts?

2.6.2 *Idealization and Abstraction*

In her seminal study Nancy Cartwright (1989) distinguishes idealization from abstraction in the natural sciences on the basis that the former, much like a caricature, distorts an object or a system to the point that it no longer describes with accuracy an object in the physical world, while the latter is a representation that omits some of the properties of the object, leaving out features that the object has in its concrete form. In both cases, just like in the case of isolation, we start with an existing object. But in economics all objects come into being through a process of reification; there are no objects to which these definitions could be applied. Idealization in economics is not the manipulation of a known object, but, rather, the creation of one that corresponds to, that embodies an idea. Mary Morgan (2012: 246) denies this difference and analogizes the objects of an economic model to a model of planetary motion as follows:

> Models in economics offer the same kind of explanatory services that planetary motion models offered to early astronomers in showing them how the elements of the universe that they observed fit together into a systematic account. Models show economists, for example, how all the elements of the Keynesian system work together, or how the multiplier and accelerator might together create economic growth for the world.

This notion of the model simply asserts without any further ado that the 'multiplier' and the 'accelerator' are the same kinds of objects as the planets. Because they are not, – planets are things, multipliers are concepts - because they do not bear any resemblance to physically perceivable or realisable objects, Morgan's equating of objects with concepts requires an ontological common denominator between the planets in the first case and the multiplier and accelerator in the second if one is to claim that models offer the same kind of explanatory services about the motions of the planets and the operations of the multiplier. This could be accomplished if it could be shown that planets and multipliers have a relevant property in common, or models have a property that will provide this explanatory service with respect to anything fed into it. In the former case, we need to know what that property might be; the latter alternative would be pure magic.

This process of reification, the process with which the multiplier is purportedly turned into the ontological equivalent of the planet Jupiter is the great hidden move of modern economics. It appears in Mills, Edgeworth, Jevons, Marshall, Samuelson, Mankiw, Arrow, Debreu and many others, but perhaps its most notorious and clearest expression is found in the following passage in Walras (1926: 71):

> Following this same procedure (that of the physic-mathematical sciences), the pure theory of economics ought to take over from experience certain type concepts, like those of exchange, supply, demand, market, capital, income, productive services and products. From these real-type concepts the pure science of economics should then abstract and define ideal-type concepts in terms of which it carries on its reasoning. The return to reality should not take place until the science is completed and then only with a view to practical applications. Thus, in an ideal market we have ideal prices which stand in an exact relation to an ideal demand and supply.[19]

How does this "tak(ing) over from experience certain type concepts" happen? How are these concepts generated out of "experience"? How does "experience" turn into the concept of exchange, supply, demand, etc.? And how do these "real-type concepts" turn into "ideal-type" concepts? These are, of course, familiar epistemological questions, that merit careful analysis, but for present purposes the point is that, whereas for the natural sciences the concepts are grounded in the empirically verifiable properties of things that are "constant and homogeneous", the social sciences build their concepts on the back of "experience". The objects of the natural sciences and the experience of the social sciences are ontologically unalike. The measurability of the former is directly derived from the measured properties of the object. Indeed, the unit of measurement contained in a concept of the natural sciences is homologous with the unit with which the properties of the object were established. But this is not the case for the social sciences. "Experience" does not yield to the metric of the concept that purportedly represents it, from which in Walrasian terms, the concept takes over. Concepts "take over" from experience not in the sense of deriving from it - as it happens in the natural sciences - the measurable properties with which the concept is formulated, but in the sense of making sense of it. We humans do this in the ordinary course of our lives by looking for some sort of a common quiddity, a common essence among the various experiences we encounter, or, in the alternative, by isolating the *propria* of our experiences and

[19] Note that, whereas there are no fewer than ten examples are given for „type-concepts", 'ideal-type concepts" are left without any.

merging the ones with a commonality into a generalisation. Both of these operations are undertaken in order to find meaning in experience, both are necessarily undertaken from our point of view, and both are operations of practical as opposed to theoretical knowledge. But neither of these processes will produce notions like marginal propensity, equilibrium, elasticity or any of the other abstractions dear to economists all of which are constructed to map causal relationships. *Pace* Keynes, these abstractions cannot be obtained through introspection.[20] The general equilibrium theory rests on fourteen assumptions of which eleven cannot be true in the world as we know it, and all the others involve at least one such assumption, namely the *ceteris paribus* clause.[21] Instead, these abstractions are *regulative ideals*, that is to say, idealisations constructed to either order or to outright generate historical experience in accordance with its parameters. As such, they are not the sorts of idealisations described by Mary Morgan, but, rather, they are the inventions of an ideal type that does not exist either in nature or in experience. They are not the product of either isolation, idealisation or abstraction in the sense in which these procedures are understood in the natural sciences, but, rather they are inventions, contrivances, products of the imagination, the tools with which a surrogate reality may be constructed to which the discipline's methodology may be applied in place of the empirical reality which resists such application. In short, the objects of economics are ideological, serving the *telos* with which economic activity is undertaken and society is organised in its pursuit. Isolation, idealisation and abstraction cannot be meaningfully performed where its object is inherently indeterminate, where, in the words of John Meynard Keynes, the material is "neither constant nor homogeneous"[22] Exchange, demand, market, capital, etc. are inventions of the mind supported, where necessary by institutional arrangements – in current terms mind-dependent – that, but for human mental activity would not exist. Importantly, they not only "take over" from experience; they generate, organise and direct it. Unlike the law of gravity that does not tell the apple how fast to fall, the marginal propensity to consume tells homo oeconomicus, if he wishes to be rated rational, how to consume.[23] If,

[20] „I want to emphasise strongly the point about economics being a moral science. I mentioned before that it deals with introspection and with values. I might have added that it deals with motives, expectations, psychological uncertainties." Letter to Roy Harrod, 10 July 1938 (misdated 16 July) in *Collected Letters of John Meynard Keynes*.

[21] This clause is usually interpreted the same way as in the natural sciences, namely that all other factors remain constant. In the natural sciences this assumption has verifiable truth value. All factors shown to affect the object under study are accounted for and there are means to hold them constant. But in economics we do not know what these other factors might be, there are no means to hold them constant, so the proper interpretation of the clause is that it is assumed that there are no other factors.

[22] This clause is usually interpreted the same way as in the natural sciences, namely that all other factors remain constant. In the natural sciences this assumption has verifiable truth value. All factors shown to affect the object under study are accounted for and there are means to hold them constant. But in economics we do not know what these other factors might be, there are no means to hold them constant, so the proper interpretation of the clause is that it is assumed that there are no other factors.

[23] Letter to Roy Harrod, ibid.

however, he does not respond to the invitation as he is wont to, having been deprived of the autonomous will with which he could accept or reject it, he is no longer admitted to the set of protagonists (all of whom are perfectly alike) about whom economic theory is formulated.

Not surprisingly, the outcome is a highly elaborate and rigorously specified methodology in search of a subject. Because a suitable subject for the methodology cannot be found in the real world, one is invented for it. In contrast with the natural sciences where the object of the study exists independently of the inquiry, economics generates its subject. The lack of a settled answer to just what is the subject matter of this putative science is a good indication of the tension between the search for scientific status and the lack of a subject matter that could be subjected to scientific investigation. Historically, the more elaborate and mathematical the discipline became, the more it lost descriptive adequacy, and the less descriptive it became, the less it could specify its subject to the point where it was finally announced that economics is a method without a subject. From the highly pragmatic "how to provide revenue or subsistence for the people and the state" (Adam Smith, 1991: 374), the search proceeded to "economics is the science which studies human behavior as a relationship between given ends and scarce means which have alternative uses" (Robbins, 1932), and then to the notion that economics is not a subject, but a method of inquiry (Gregory Mankiw) or the idea that it is a specialized form of rhetoric (McCloskey 1985). But a methodology without a subject or a rhetoric with an ill-defined one cannot be the subject of law-like generalizations. Drawing parallels between 'human behavior' and, say, the elements set out in the Periodic Table, on the basis of an allegedly shared methodology is the ultimate expression of the epistemic fallacy.

Economics then is not a science not because of its modest predictive or explanatory record, – a defect that could be corrected if a suitable ontology could be had - but because its subject matter is ontologically unsuited for theoretical knowledge of universal applicability. Its subject matter does not meet the Thomist *adequatio ad rem* requirement reaffirmed by Hilary Putnam (1975: 73). The structure of ideas about things should mirror the structure of the thing about which ideas are formed. In Putnam's (1975: 69) reformulation:

> A realist (with respect to a given theory or discourse) holds that (1) the sentences of that theory are true or false; and (2) that what makes them true or false is something external – that is to say, it is not (in general) our sense data, actual or potential, or the structure of our minds, or our language, etc.

But economics cannot meet this requirement because its subject matter is not independent of its theories. There is no *res* with its own structure about which ideas mirroring that structure may be formed. Its objects are abstract entities created by theory and the structure of these abstract entities is infected by the requirements of theory that both generates these entities and theorises about them.

Modern economics is an ideology presenting itself in scientific garment, but, in fact, it is promoting a particular agenda. Although its prescriptions are presented in a form that mimics the form of the laws of the natural sciences, it is concerned with identifying the sort of behaviour that is most conducive to achieving its ideological ends, such as maximising efficiency, understood and measured as the return on capital employed. The choice of efficiency as the foundational value and its measurement as the return on financial capital employed, may or may be a "rational", a desirable or laudable choice, but it is a normative choice, and not a science. Most importantly, its purported moral neutrality – obtained by the claim to scientific status - is a dangerous and destructive deception.

The undoubted wealth enhancing achievements[24] of the effort to reconstitute economics so as to make it suitable to this simulated, in Hausman's words "inexact and separate science", have come at a price. First, it has altered our lives by constructing roles, fostering laws, rules and institutions that comply with its claims. In the process the human being has been turned into the sort of determined passive object, that is the typical subject matter of the natural sciences. It has diminished our moral competence by curtailing our autonomy in the name of efficiency. The morally competent human being, whose life revolves around the resolution of incommensurables requiring judgment, is being crowded out by the utility maximising rational choice automaton. He and she have been turned into the instrument of a theory. Second, it has displaced the question that should be the proper subject of economics, – to wit, how do we and how should we choose among incommensurables - with a fatuous axiom about a greater gain being preferred over a smaller one. Economic life is not about equilibria, marginal propensities, Pareto optima or rational choice, none of which exists except in human conjecture, but the exact opposite of these; how do we make justifiable decisions about incommensurables in a world of uncertainty, violence, unending, in many respects growing wealth and income difference, alienation and injustice. Economics proposes that the decisions be made in accordance with Mill's axiom, to wit, in the pursuit of material gain, but the choice of that objective, whatever its legitimacy or desirability is just that: a choice from among many other possible choices in the search for human happiness and fulfilment. Despite its claim to the contrary, the subject of economics is much less about what *is* – a task that could be better met by greater descriptive adequacy than by assumption based theorising – and much more about what *ought to be*. Its models are allegories, constructions of surrogate realities designed to set out the consequences of idealised assumptions with the undertlying message that the task of public policy is to bend social reality towards those idealised assumptions. As such, they can be very useful and instructive tales guiding us in our choices, but they are not scientific instruments. They can also be very harmful because in limiting itself to the mathematically representable, models can convey a world in which the elements of reality that are incommensurable with the model's axioms are left out of account, and, having been left out, they are either allowed to perish, or are quite deliberately killed off.

---

[24] How much of this achievent is due to science and how much of it to economic theory may be the subject of an interminable debate. But there can be little doubt about the subtly corrupting effect on science of having been steered towards the study of subjects that enhance 'efficiency' and 'consumer satisfaction' as those notions are set out in modern economic theory.

Accordingly, modern economics has been most convincing when it has paid attention to the problems caused by the lack of homogeneity in the reality it examines. Studies addressing small, spatially and temporally well defined issues offering a generous supply of empirically verifiable evidence (Dasgupta 2002) have been more effective in guiding us as to what should be than those aiming for the formation of law-like generalisations of universal applicability. The modern method of analysing a statistical population, where the "population" consists of some conceptual unit but containing heterogeneous objects seems to be a more promising foundation for economics than the universalising system building of modern economics. It is a method that studies recurrences within a spacially and temporally defined "population" without privileging any value or property and without attempting to contrive universals, and, in doing so, it permits an answer to the question: what is to be done? Without building into its system the values with which the question is answered.

References

Arendt, H. 1951. *The origins of totalitarianism*. Penguin Classics Edition (2017).
Ayer, A.J. 1946. *Language, truth and logic*. New York: Dover Press.
———.1956. *The Problem of Knowledge*. London: Macmillan.
———. 1963. *The concept of a person*. New York: St. Martin's Press.
Cartwright, N. 1989. *Nature's capacities and their measurement*. UK: Oxford University Press.
Dasgupta, P. 2002. Modern economics and its critics. In *Fact and fiction in economics*, ed. U. Mäki. Cambridge: Cambridge University Press.
Friedman, M. 1953. *Essays In Positive Economics*. Chicago: The University of Chicago Press.
Gibbard, A., and H. Varian. 1978. Economic models. *The Journal of Philosophy* 75 (11): 664–677.
Hamminga, B., and N. DeMarchi. 1994. *Idealisation in economics*. Amsterdam: Rodopi Press.
Hausman, D.M. 1992. *The inexact and separate science of economics*. Cambridge: Cambridge University Press.
Helmholtz, H. 1887. *Counting and measuring*. New York: D. Van Nostrand Co.
Jevons, W.S. 1911. *The Theory of Political Economy*. Fourth, (reprinted 2013). UK: Palgrave MacMillan.
Jovanovic, F., and P. LeGall. 2001. March to numbers: The statistical style of lucien March. In *The Age of Economic Measurement*, ed. J. Klein and M.S. Morgan. Durham: Duke University Press.
Kaldor, M. 1974. What is wrong with economics? In *Further essays in economic theory*. (1978). New York: Holmes & Meier Publishers, Inc.
Kline, M. 1980. *Mathematics: The loss of certainty*. Oxford: Oxford University Press.
Mäki, U. 2002. Kinds of assumptions and their truth: Shaking an untwisted F-twist. *Kyklos* 53 (3).
———., ed. 2009. *The methodology of positive economics: Reflections on the Milton Friedman legacy*. Cambridge: Cambridge University Press.
McCloskey, D.N. 1985. *The rhetoric of economics*. Madison: The University of Wisconsin Press.
Mill, J.S. 1843. *System of logic*. Longmans: Green & Co. London.
Morgan, M.S. 2012. *The world in the model: How economists work and think*. Cambridge: Cambridge University Press.
Musgrave, A. 1991. Unreal assumptions in economic theory and their truth: The F-twist untwisted. *Kyklos* 34 (3).
Nagel, T. 2016. *What is it Like to Be a Bat?* Stuttgart: Reclam.

Nussbaum, M. 2001. *The fragility of goodness*. Cambridge: Cambridge University Press.
Polanyi, M. 1958. *Personal knowledge*. London: Routledge & Keegan Paul.
Poovey, M. 1998. *A history of the modern fact: Problems of knowledge in the sciences of wealth and society*. Choicago: The University of Chicago Press.
Putnam, H. 1975. *Philosophical papers*. Vol. 2. Cambridge, MA: Harvard University Press.
Putnam, H. 2002. *The collapse of the fact/value Dichotomy and other essays*. US: Harvard University Press.
Putnam, H., and V. Walsh. 2012. *The end of value-free economics*. Abingdon: Routledge.
Quine, W.V. (1951/1976). *The Ways of Paradox and other Essays*. US: Harvard University Press.
Robinson, J. 1981. *What are the Questions? And Other Essays*. New York: M:E: Sharpe Inc.
Russell, B. 1919. *Introduction to Mathematical Philosophy*. (2008) ed. Spokesman.
Smith, A. 1776. *The wealth of nations*. New York: Alfred A. Knopf. (1991).
Stewart, D. 1792. *Elements of the philosophy of the human mind in the collected works of Dugald Stewart*. Vol. 1. Edinburgh: T. Constable and Co. 1854.
Walras, L. 1926 *Elements of pure economics*. Trans. W. Jaffe (1954). George Allen and Unwin Ltd. London.

Chapter 3
Positioning and the Nature of Social Objects

Stephen Pratten

Abstract In opposition to the ontological neglect that characterizes so much economics a group of researchers based in Cambridge in the UK argue that method and substantive theory can benefit if informed by explicit, systematic and sustained social ontology. This paper sets out key elements of the general socio philosophical ontology that this Cambridge group outline and explores how within this approach social reality is understood as being constituted through the emergence and transformation of social totalities. A particular focus is on the importance placed by the Cambridge approach on processes of social positioning and in exploring what this involves. The paper highlights that it is this general socio philosophical ontology that provides the basis for the critique of the dominant methods of modern economics since these presuppose an implicit ontology quite inconsistent with both it and indeed any realistic conception. This implicit ontology is also shown to constrain the way economic objects are characterized within mainstream economics. Finally the paper provides an illustration of how powerful accounts of specific social existents can be developed that are consistent with the broader social ontological conception that the Cambridge group defend by sketching one account of the firm and the modern corporation recently forwarded that draws systematically upon the notion of social positioning.

3.1 Introduction

In recent years a group of researchers based in Cambridge whose work is sometimes identified as *Cambridge Social Ontology* have developed a distinctive critique of modern economics that traces the widely recognized explanatory failures and lack of realisticness of much of the contemporary discipline ultimately to a pervasive ontological neglect (see Lawson 1997, 2003, 2015b and Pratten 2015 and Faulkner et al. 2017). According to this critique theories and methods of modern economics

S. Pratten (✉)
King's Business School, King's College London, London, UK
e-mail: stephen.pratten@kcl.ac.uk

© Springer International Publishing AG, part of Springer Nature 2018
P. Róna, L. Zsolnai (eds.), *Economic Objects and the Objects of Economics*, Virtues and Economics 3, https://doi.org/10.1007/978-3-319-94529-3_3

are typically fashioned in ways that pay scarcely any attention to the nature of the social material to which they are applied, i.e., within mainstream economics the characterization of economic objects, the choice of method and the development of theory proceed without any sustained consideration of the sort of stuff that is being addressed. The assessment advanced by this group is that economists have been especially concerned to be seen to be conforming to proper scientific method and in doing so have drawn upon a conception of scientific method that is erroneous. Part of the work of this group is devoted to showing how the neglect of basic ontological issues and the accompanying scientistic orientation has led directly to the widespread failings of modern economics.

In opposition to the ontological neglect that characterizes so much economics this Cambridge group argue that method and substantive theory can benefit if informed by explicit, systematic and sustained social ontology. They sketch a constructive program where advances in method, substantive theory and ontology are seen as being best produced together with developments in each informing the others.

A central preoccupation of the Cambridge group has been with the elaboration of the broad principles of social reality constitution that is with theorizing the nature of social being in general. The general social philosophical ontology that they outline turns on a set of distinctive core principles or assessments. A further contribution of the Cambridge group has involved them drawing on this general social philosophical ontology to theorize specific social existents such as money (Lawson 2016a), gender (Lawson 2007), technology (see Faulkner and Runde 2013 and C. Lawson 2008 and 2017), trust (Lawson 2001 and Pratten 2017), institutions (Lawson 2015a and Pratten 2016) and the corporation (see Lawson 2015c, d and Deakin 2017). Central to all these contributions is the conception of social positioning by which people and social phenomena generally become positioned as components of emergent social systems or totalities.

In this paper I set out key elements of the general socio philosophical ontology that the Cambridge group outline and explore how within this approach social reality is understood as being constituted through the emergence and transformation of social totalities. A particular focus is on the importance placed by the Cambridge approach on processes of social positioning and in exploring what this involves. The paper highlights that it is this general philosophical social ontology that provides the basis for the critique of the dominant methods of modern economics since these can be shown to presuppose an implicit ontology quite inconsistent with both it and indeed any realistic conception. This implicit ontology presupposed by the methods insisted upon is also shown to constrain the way economic objects are characterized within mainstream economics. A final objective of the paper is to provide an illustration of how powerful accounts of specific social existents can be developed that are consistent with the broader social ontological conception that the Cambridge group defend by sketching one account of the firm and the modern corporation recently forwarded that draws systematically upon the notion of social positioning.

3.2 Cambridge Social Ontology

The general social ontology set out and defended by the Cambridge group turns on certain core principles. Here I provide the briefest summary of certain relevant elements.

For the Cambridge group the category of social reality denotes the set or totality of all phenomena whose existence *necessarily* depends on human interaction. A starting point here is an assessment that social reality is constituted in large part by *emergent totalities*, where people and things broadly conceived become incorporated as components.

A central category within this approach is that of emergence and in particular the emergence of social totalities. The term emergence is taken to relate to the appearance of novelty or something previously absent or unprecedented. This characterization of emergence emphasizes the role of *relational organization* and recognizes the operation of both efficient and formal causality. A totality is understood as a system of organized, usually more basic, or pre-existing, elements that reveals a coherence or integrity at the system level. Totalities emerge through the relational organization, perhaps with some modification, of such pre-existing elements, the latter being effectively harnessed and organized as components. How the elements are organized makes a difference; it is due to their organization that the totality has properties that extend beyond those of the individual elements. The influence that relational organization has is that of formal causality. Some of the elements will be organized so that they are tied to an environment in some way. Thus emergence always occurs in a context. Since an environment appears along with any totality, it too is an emergent. In the case of social totalities, the elements so organized either may or may not themselves be socially constituted.

Systems or totalities that emerge in this manner, since they are constituted in part by the way in which the components are arranged or relationally organized, are irreducible to the sum total of its isolated components. So on this view ontological reduction of a dis-aggregative sort is seen as being proscribed. Along similar lines any emergent powers of efficient causation of emergent totalities are causally irreducible in that they are not predictable from the causal powers of these elements when considered in isolation. Thus causal reduction of any simple sort is also invalid.

As a way of illustrating some of these points certain everyday examples can usefully be drawn upon. Thus consider a house and its construction:

> The components include bricks, mortar, wood, panes of glass, cement, etc. Of course there will be a context, a plot of land, and this will be prepared so that the various components can relate to it in an appropriate manner. At any stage in the process of construction an observer will find not only the part of the building constructed so far, formed out of various components, but also the relational organization of the latter components (to each other, and to the elements in their environment). And this organization will be essential to the house's construction and properties. As the house is completed, so is the relational organization of the house's components; the two – the totality and the organizational structure - emerge simultaneously. Each are causal, but in different ways. The house has power to provide

safety and shelter to facilitate family and other indoor activities, to be bought and sold and so on. The arrangement of the parts makes the house feasible. The latter is a form of formal causation (Lawson 2016c: 363).

In those cases where an emergent totality includes people amongst the components then, within this Cambridge perspective, it is understood as being a *community*.[1] Meanwhile the term *artefact* is used to refer to those totalities where there are no human components and it is recognized that the components themselves are likely to be artefacts of one form or another.

The term "social positioning" captures within this Cambridge approach the process whereby, through general acceptance by members of a community, human individuals, things or other phenomena become incorporated as components of emergent totalities. Social positioning is seen as centrally involving the generalized acceptance of three elements in regard to any item that is thereby positioned. First, the practical placement of the item within the structure of a (already established or possibly emerging) system or totality takes place. Second, certain capacities of the item are harnessed to serve as one or more system functions of the totality. In the cases where it is human beings that are being positioned, their capacities are harnessed via the allocation or ascription of certain positional rights and obligations. Third, the item in question is (usually) allocated a status or distinct identity.

As a way of illustrating this positioning process consider how human beings in a university are positioned as lecturers, students, administrators, librarians, cleaners etc. Each case of positioning involves both the acquisition of a status or identity (of administrator, lecturer, student, etc.,) and also a practical placement as a component within the teaching and or research structure of the university. This process involves the acquisition of rights and obligations with each right (obligation) of one position being linked to an obligation (right) tied to a different position. The right of students to attend lectures and tutorials is matched to the obligation of lecturers to prepare and provide them. The right of both lecturers and students to access and use specific lecture theatres is matched by the obligation those making up the university administration have to fund, service and maintain them. These matching rights and obligations are both social relations and power relations. It is positions, rights and obligations and the *collective practices* (see below) associated with them that provide the organizing structure of the communities of which they are a part. In the case of the university, the rights and obligations acquired are such as to ensure that the positioned participants can and do interact in ways that serve the university's aim of dispensing education and facilitating research.

Turning from communities to artefacts the positioning process in this context can be illustrated by noting how panes of glass may be positioned (identified and practically placed) as windows in a house, and blocks of wood as doors. In this process certain capacities of the positioned objects get to serve as *system functions*. Thus, from the various capacities that each pane of glass has it is that of allowing light

[1] See Pratten (2013) for some reflections on how the characterization of the community by the Cambridge group compares with certain alternative perspectives.

through whilst withstanding the weather that is picked out when it is positioned as a window and it is this capacity that comes to serve as the pane's system function.

The establishment of a given position as well as the allocation of people or objects to it are ultimately understood within this approach as a matter of community acceptance. According to this Cambridge perspective it is important to recognize that by acceptance there is no suggestion of agreement. Acceptance may only imply a begrudging willingness to comply or a hesitant preparedness to go along with. It is recognized that the establishment of either a position or the determination of who or what should occupy it can come about by a legitimate community authority in effect declaring something to be so or in an altogether more spontaneous manner.

An especially important category within this Cambridge approach is that of collective practice. To be a practice is to be a way of proceeding or going on that is actually pursued or followed. Here the emphasis is on both a way of proceeding as well as on the fact of it actually being followed. According to this perspective to be engaged in the activity of say eating an apple or running for a bus is not yet to be involved in following a practice. If an individual regularly peels an apple in a certain set manner before eating it, this may constitute a practice, a *way* of eating the apple, or at least, preparing it for consumption.

The intention here is not for the term collective practice to express something that will somehow be carried through other than by individuals (or groups of individuals). It is collective not in relation to who performs it but in the sense that within the relevant community it is an accepted or done way for an individual or group to proceed to reach a particular end in a given context. Effectively it is a practice that has been positioned. As such it is a practice that has the status of being accepted as legitimate in the given circumstances within the community. From this perspective a collective practice is an accepted practice in the sense of being recognized or acknowledged as a currently done way within a specific community. In order to get by, in order to participate at all, in a particular community individuals generally adhere to the relevant collective practices.

Acceptance in this context is essentially a form of collective or community recognition. The idea of individual practices being accepted in the relevant sense is in effect a status. It is community wide recognition that grounds the collective practices that serve to signal ways of proceeding as being the done ways of proceeding within a community. A collective practice then being an accepted way of proceeding within a community is a way of proceeding with which each community participant typically complies recognizing (if often only implicitly) that this constitutes a condition of community participation.

The emergence of a specific collective practice is consistent with a certain amount of variety at the level of individual practices in a relevant domain. It is at this point where the notion of a norm enters the picture. A norm, within this Cambridge approach, is understood as any component or aspect of a collective practice that is compulsory and thus that ought not to be deviated from. So interpreted a norm might be reformulated as a segment or an aspect of a collective practice that carries within itself, if typically implicitly, an obligation. But understood in this manner a norm remains a material component of a collective practice. And importantly, a norm can exist implicitly without notions of obligation being linguistically articulated.[2]

Where a set of ways of proceeding are constituted as examples of a collective practice, as community-accepted ways of proceeding, this involves the participation of effectively all members of the community. The fundamental point or function of collective practices *qua* collective practices, according to the Cambridge group, is the coordination of community life. They are an essential part of the organizing structure of all totalities that are communities facilitating the coordination of the separate activities and goals of individual participants. The organizing structure can work in this way only to the extent that there is a broad adherence to the community accepted ways of proceeding. So collective practices, in being constituted as ways in which things are in fact done in a community in specific contexts are by that token also normative in nature. By indicating how things are done, they at the same time inform potential community participants how things should be done.

Within this framework to be a community participant is to enjoy the rights of participation as well as the obligations that go with it. Each participant anticipates that other participants will conform to the obligations implicit in accepted collective practices, and has the right to object, issue rebukes, and so undermine individual behaviors interpreted as deviant (relative to those collective practices). Such objections and rebukes reflect at a concrete level the broader community requirement that currently accepted collective practices be conformed with. To the extent that it is everywhere structured through accepted collective practices social being is continuously subject to the wielding of social systematic power by all of us over each other.

According to this Cambridge approach the positioning process presupposes a background of trust. It is the exercise of the human capacities of trusting and being trustworthy that ensures that human society does not fall apart and where these capacities are not sufficiently exercised integrity at the system level is threatened. Trust and the social positioning process are intimately connected and fundamental in this account of social reality:

> Notice that the role of rights and obligations in structuring social life presupposes the human capacities of being able to be trustworthy and to be trusting of others, of being willing and able both to make and to keep promises and other commitments, and to believe that others can and will do so. It should be clear that these human capacities are necessary conditions for the interactions involved to occur, for obligations in particular to be efficacious. As such these capacities of trusting and being trustworthy, etc., qualify for being considered as the glue of social reality, as the adhesive that enables the organizational structure to achieve a degree of binding (Lawson 2014: 27).[3]

---

[2] According to Lawson within linguistic communities norms can always be codified and where they are rules are thereby established. For further detail on his account of norms and rules, see Lawson (2016b).

[3] For extended outlines of the general socio philosophical ontology that the Cambridge group defend see Lawson, 2003, 2012 and Pratten 2015 and 2017.

3.3 Ontological Neglect within Mainstream Economics and its Consequences

The priority given by the Cambridge group to sustained ontological analysis has facilitated the development of a distinctive critique of mainstream economics. It is shown that mainstream economists are committed to a particular method and view the method as essential to proper scientific endeavor. Mainstream economists take it for granted that mathematical methods are essential to the doing of appropriately scientific research, often supposing that the ultimate goal is some form of event prediction. A fundamental insight of the Cambridge group is that, for any method to be able to illuminate a domain of reality, the nature of the phenomena of that domain must be of a sort to render this feasible. From the perspective of the Cambridge group there is a sense in which method must fit with the nature of its object. If we set ourselves the task of cleaning a window, we do not choose an industrial drill or a big stick however useful these may be in other contexts. We know enough about the nature of the glass and the properties of the drill and the stick to recognize that any such choice would be inappropriate – a soft cloth is chosen since it fits better with what we know about the glass. For a research method to be relevant in a specific context the material to be studied must be of a sort that makes feasible the method's deployment. Rather than starting with a question about an aspect of social reality and determining an appropriate method, modern mainstream economists typically start with a particular type of method and erroneously presume it is appropriate to all social contexts. A key feature of mainstream economics, according to the Cambridge group, is its profound ontological neglect and general philosophical naiveté. The result of this pervasive ontological neglect is that modern mainstream economists end up distorting social phenomena and mischaracterize economic objects so their favored methods can be applied.

Mainstream economists hope to be scientific in the sense of natural science. However, as the Cambridge group show, they are influenced in this by a misconception of science. If science is the move from phenomena of interest to underlying causes then this move is as feasible in economics as in natural science. In some special cases natural scientists can successfully carry out controlled experimentation. In such conditions, event regularities may be produced in the course of their evaluating a theory about an underlying causal mechanism. Mainstream economists, though, have typically not only restricted their focus to this special case, but also universalized the wrong aspect of it. Instead of seeing the central moment as furthering our understanding of some underlying causal mechanism, economists have focused on the experimental production of event regularities. Moreover, they have generalized this inessential feature of science to a context, namely the social realm, where typically the conditions are such that event regularities do not occur. Thus, although a science of social phenomena remains entirely feasible, prediction and the reliance on methods of mathematical modeling in economics, presupposing as they do a ubiquity of event regularities, constitute significant obstacles preventing the discipline from realizing its full scientific potential.

By carefully considering experimentally produced event regularities the Cambridge group identify three conditions that must be satisfied if event regularities of the sort presupposed by the view of science adopted by mainstream economists are to be guaranteed. Ultimately in order for an event regularity to be guaranteed the relevant domain of reality must consist of factors that are: (1) intrinsically stable, (2) isolatable and (3) actually acting in a condition of isolation. If an event regularity is to be achieved then isolation at some level is required. In the relevant philosophy of science literature a situation in which many and changing causal mechanisms determine the course of events is referred to as open, whereas one in which a single mechanism is isolated and an event regularity produced, is usually described as closed.

The Cambridge group, through sustained ontological reflection, suggest the closure conditions (i.e., the conditions required to guarantee that event regularities obtain) are unlikely often to be met in the social domain and thereby clarify why mainstream economics operating with a conception of science as hinging on successful prediction and necessitating the use of mathematical methods does so poorly.

An initial observation made is that even if economic forces were intrinsically stable and isolatable there can be no presumption that they act in isolation. Even in the natural realm strict event regularities are not the norm, experimental intervention for example is required to achieve conditions of isolation. Thus in seeking to apply their modeling methods to all social situations mainstream economists ignore the special conditions required for such regularities even within natural science.

The problems facing those who presuppose that event regularities are a prevalent feature of the social domain are shown by the Cambridge group to be considerably more substantial. Rarely is it the case that social phenomena are either intrinsically stable in the way that some natural mechanisms seem to be or isolatable. With regard intrinsic stability the Cambridge group note that more or less by definition social phenomena depend on us for their existence. Without us there would still be natural forces like gravity, but not social phenomena like tables, chairs, cars, markets, universities, language systems and so on. By considering the last example more closely the problems associated with any assumption that the social domain is constituted by systems with sufficient stability for event regularities to be guaranteed can be appreciated. A language system is given to us at any moment and facilitates our speech acts. But it also depends on us. By drawing on it when speaking, individuals, in total come to reproduce the system in question. In part we also transform it; whether intentionally or through error we continually change aspects of it. The language system is best understood as something that is being continually reproduced and transformed. This is its mode of being: it exists as a process of reproduction and transformation. A language system is thus intrinsically dynamic in nature. All other social systems – firms, markets, universities, public agencies, etc., – are best understood in similarly dynamic terms. From the perspective developed by the Cambridge group all social phenomena are processual in nature. While it is acknowledged that aspects of social life may be reproduced over sometimes significant regions of time and space, the sort of stability that may be attributable to say gravitational forces on the surface of the earth is unlikely to prevail.

The Cambridge group argue that the further requirement needing to be secured in order to guarantee an event regularity, namely that social causes be isolatable, is at least as unlikely to be satisfied, the social realm seems intrinsically open and interconnected. Firms, markets, institutions, social relations, even individual identities, cannot be experimentally isolated from each other. Indeed most if not all social phenomena are actually constituted in their relations to something else. So employers exist in relation to their employees, similarly landlords cannot exist without tenants or at least potential tenants, teachers without students, parents without children and so on. As we move into new positions as students, employees, registered nurses, trade union members etc., what we can and cannot do is determined by our relations to others.

The emphasis of mainstream economists on using mathematical formalism likely reflects their desire to be scientific in the sense of natural scientists. This is understandable, perhaps even laudable, in that the latter do regularly achieve success in advancing human understanding. A less benign interpretation might see them as seeking to pose as scientists merely in the hope of acquiring a certain status in society. Whichever of these or related motives is most pervasive in modern economics, the conception of science that mainstream economists adopt is mistaken. The Cambridge group argue that a causal conception of science is far more tenable. They suggest that if there is a defining moment or feature of science, a more sustainable contender is causal explanation, where causal explanation is itself given a particular meaning. Causal explanation is understood as relating to the move from a conception of some event or phenomenon of interest to the identification and understanding of an underlying causal factor that (in part at least) produced it. It is the move from, say, symptoms of an illness to the underlying cause making the difference. The Cambridge group emphasize that causal explanation so understood is a step that typically cannot be reduced to mathematical manipulation. It involves a conceptual move from phenomena of one type to a cause of a quite different (and perhaps previously unknown) type. If the latter conception is accepted then economics and other social sciences can indeed take their place amongst the sciences in the sense of natural science, though, given the nature of social material, the current emphasis on prediction and formalism obstructs this outcome. The insistence by mainstream economists on specific methods prior to evaluation of their relevance encourages social systems to be characterized in highly distorted ways. It is this primary insistence on certain methods that the Cambridge group argue will have to be abandoned if a science of, or indeed any explanatorily useful, economics is to emerge.

3.4 The Firm and the Corporation as Positioned Communities

Whilst the Cambridge group have focused over a considerable period on developing the broad socio-philosophical ontology sketched above and exploring its implications for all aspects of social theorizing, in recent years there has also been a move to draw upon it in order to theorize specific social existents including money, gender, trust, technology, institutions, the firm and the corporation. In contrast to mainstream economics where an implicit ontology implied by a commitment to particular methods, that in error are understood as being essential for science, encourages the distortion of social reality and mischaracterization of social objects, the Cambridge group in developing understandings of specific social phenomena build on an explicit and systematized social ontology. This is not to suggest that compelling accounts of particular social existents can be simply read off from the general social ontology defended, additional empirical input is required. Here I will provide an illustration of how powerful accounts of specific social existents can be developed that are consistent with the broader social ontological conception that the Cambridge group defend by considering the community based account of the firm and the corporation set out by Tony Lawson in certain recent contributions (2012; 2015c, d).

Lawson's accounts of the firm and the corporation very much draw on the insights regarding the nature of social reality reviewed previously. Thus society is seen as being made up in large part of *communities.* Communities are viewed as pervasive within the social realm. Communities are structured social entities that emerge from human interaction and include human individuals amongst their components. It is positions and associated rights, obligations and collective practices that form the structure of communities and relationally organize the human components. All social organizational forms, all social entities that include human beings as components, are forms of community and all are organized through relations of rights and obligations that rest on the exercise of human capacities of trusting and being trustworthy. On Lawson's account both the firm and the corporation are fundamentally forms of community. But what kind of community is a firm and what kind of community is a corporation? Lawson argues that certain communities take on the identities of a firm or a corporation by being appropriately *positioned.*

In order to appreciate the argument it is necessary to recall how persons and artefacts are respectively seen as acquiring identities. Let us first consider the case of artefacts. When an artefact is positioned certain of its capacities become interpreted as its positional functions:

> objects become positioned as say tables, seats, eating and drinking and serving utensils, etc. Over different times, places and cultures, the objects so positioned will vary in shape size, form and material content (just as, in any point in time and space, there are very often entirely different inanimate objects that could have been successfully positioned in place of each). But in all cases, positioned objects of the sort listed facilitate the needs of a system of human beings participating in collective practices bound up with sharing food together. And the (set of) power(s) of each of the objects that contributes to this end is seen as its (set of) function(s). (In turn of course the meal itself may be a component of a wider system, functioning perhaps to facilitate regular family or tribal gatherings, or in specific cases perhaps to celebrate a family member's birthday or mark another occasion.) (Lawson 2012: 376).

In contrast, when a human individual is positioned as, for example, an army officer, a midwife or member of a jury, the individual becomes the bearer or agent of novel positional powers and specifically rights and obligations. Positions and positional rights and obligations are typically in place in order that certain perceived needs of the system can be met by appropriately allocated occupants. So the positioning of human beings is typically also functional. But in the case of the positioned human individual it is rights and obligations that are acquired.

How does this bear upon the nature of the firm? The account of the firm and the corporation developed by Lawson has at its centre the idea that it is not only artefacts and people that can be positioned and thereby acquire identities but so too can communities. The firm on this view is a community positioned as a component of a wider industrial community. In being so positioned the community that is positioned as a firm goes through a process of positioning that is analogous to that associated with artefacts – a set of the community's causal capacities become fairly straightforwardly interpreted as the firms wider system functions. Generally recognized examples of firms in the UK include sole trader, business partnership and limited company. On the account of the firm set out by Lawson, conditions that need to be in place in order for a community to become a member of a particular firm type are: (i) it has to be appropriately legally positioned or registered, (ii) it has to possess a legal structure associated with the relevant position, and (iii) all such legal structures presuppose a conception of the set of capacities the candidate community must possess and which, on positioning, are to constitute the community's characteristic system function.

Building on this analysis of the conditions necessary for the establishment of a firm Lawson proceeds to provide the following general characterization:

> The (modern) firm is simply a specific community, currently legally positioned, that is formally registered, within the wider, typically national (or international) community, as an emergent sub-community of the latter, oriented to the collectively coordinated production of goods and/or services to be sold to others, in a way that is intended to be advantageous to (at least some of) the community members. It is normally the case that (at least some of) that advantage is interpreted as 'profit'. (Lawson 2015c: 15).

A significant advantage of the kind of approach Lawson develops is that it facilitates a clear articulation of the distinctive features of the corporation as opposed to the ordinary unincorporated business firm. To appreciate the issues here an understanding of how multiple processes of positioning unfold is necessary. Both artefacts and people can be positioned in various systems:

> Very often it is the case that certain items can be and are positioned in two or more different systems simultaneously and so possess two or more sets of functions, one for each system (just as a human individual can simultaneously serve as say a parent and a grocer). Thus an item of clothing may serve both as part of a system providing protection from the elements for a particular human being in winter, say, and also by identifying the wearer of the clothing as a police person, members of a specific football team or a bride or a priest, as part of an additional more collective system. A house or painting or ornament may function according to the specific uses made of it in servicing, say, a family home, but share in a common a power to retain, and so function as a store of value in another system. Similarly, of course, those powers or properties interpreted as functions can change as objects are repositioned. An item for sale in a shop is identified as a commodity and has the function of being tradable for credit at a given price. But from the perspective of a different system and once purchased it can be inserted into a system where its function is in line with its more specific uses (as a hammer, screwdriver, etc). (Lawson 2012: 376-7).

It is useful to distinguish various forms of multiple positioning. Thus, the examples of a specific individual being positioned as a parent and a grocer illustrate a process of *multiple horizontal positioning* where the same individual is positioned in different systems. It is also relevant to consider multiple vertical positioning where each of a set of positions is nested in others – an individual might be a member of a political party, a member of parliament and a member of the front bench team.

Returning to the case of the corporation – Lawson sees the registration of a community as a limited company as an example of multiple vertical positioning applied to the community in question, but significantly argues that the process is a *hybrid* of those that are involved with the positioning of artefacts and human individuals. When a community is first positioned as an ordinary unincorporated firm the positional properties that were implicated are functions. Specifically, the community's capacity of coordinated production of goods and or services for sale to others at an advantage to its members becomes its official or legally constituted characteristic system function. However, when the same community is further positioned as a limited company or corporation, the positional properties that are thereby primarily implicated are rights and obligations. As a consequence of this process of additional positioning the firm acquires particular rights and obligations in a way that corresponds to those cases where it is human beings that are being positioned. Modern companies for example have the right to own various assets, make contracts, sue and they can themselves be sued. It is in this sense meaningful to talk of the company as acquiring legal personhood.

Various features of Lawson's community based account of the corporation are worth considering in further detail. This approach helps to clarify that while the modern corporation may involve the deployment of legal fictions, this does not imply that the corporation is in any way an ontological fiction. Legal fiction is a false claim about a pertaining situation to which the legal process resorts merely in order to adjust the incidence of a pre-existing set of rules. There is nothing in the use of legal fictions that implies that the modern corporation is anything other than a real social totality with rather extraordinary powers. A corporation is simply a real community positioned under law as if it were a person.

As a community is positioned as a corporation, a separation is established between rights and obligations that accrue to individual members of the community and those that accrue to the corporation itself, viewed as a legal person. A distinctive feature of this further process of positioning, Lawson argues, is that it allows the corporation as a whole and the individual members to be substantially protected from the misdemeanors, financial failures and particular limitations of the other.

Lawson's account of the corporation facilitates an understanding of certain characteristic and problematic features of the modern corporation. Any corporation has a legal status that is separate from those of its individual members. Moreover any corporation can establish numerous effectively separate entities, many perhaps located in different countries, they can legally own the latter's shares and constitute each as partly or wholly owned subsidiaries with a separate legal status. This opens up opportunities for the endless transference of liabilities. A subsidiary can sue and be sued separately from its parent and the obligations it acquires will not automatically become obligations of its parent.

Lawson notes that the hybrid process of positioning process that characterizes the corporation can constrain regulatory attempts to control its activities. For instance in attempting to exert some degree of control, criminal courts have found that corporations cannot be held liable for criminal acts requiring intent because they do not have 'a state of mind'. Meanwhile, corporations, when it suits their purposes, can assert the right to freedom of expression, the right to a fair trial and they cannot be forced to incriminate themselves. In those cases where a corporation is found liable for something, it cannot be imprisoned or set to work in the community. The punishment imposed is typically a fine, oftentimes involving a sum the corporation concerned may have anticipated as a possibility and factored into its planning.

A feature of the contemporary corporate scene that Lawson highlights is that often there seems no one who is in control of the development of corporations and their initiatives. Specifically he suggests that the incentives are for a corporation (or its set of directors) or one of its subsidiaries, to take manifestly irresponsible, damaging and dangerous risks where the possible financial benefits are significant. He argues that there is no effective agency that feels or acts as if it can easily assume responsibility for the corporation's actions. Shareholders limited in their liability if things go wrong and who have little control over what goes on within the corporation or is undertaken in its name do not view themselves as being responsible. While it is the board of directors who do have the task of managing day-to-day operations, the dominant modern interpretation of the law has it that no matter how responsible specific individuals might be, the duty of those directing the day to day affairs of the corporation is to serve the interests of the shareholders. The current state of affairs is one where neither the directors nor the shareholders nor any other body feel they have much influence over the nature of the overall objectives of the corporation. The result Lawson concludes is a dangerous power vacuum.

A further and final feature of Lawson's account of the corporation that it is worth noting here is that it not only highlights the distinction between ordinary unincorporated business firms and corporations but serves to focus attention on how profit oriented communities came to be accepted as the sort of communities that could obtain the status of corporations. Although taken for granted today, historically there has been an enormous amount of contestation and debate over the types of communities that could be positioned as legal persons and the specific rights and obligations that could be acquired for those that were so positioned as well as regarding which agencies were to execute relevant decisions concerning the accessing of such rights and obligations.

3.5 Conclusion

Mainstream economics by adopting a particular and flawed conception of science ends up distorting social reality so as to ensure that the methods erroneously assumed to be essential for science can be deployed at all. The Cambridge group has sought to highlight the importance of ontological analysis. They have forwarded a general socio philosophical ontology that can effectively accommodate all specific social phenomena. The implicit ontology presupposed by the methods prioritized by mainstream economists can be shown to be impoverished in comparison. Once a more adequate account of science is adopted the general social ontology set out and defended by the Cambridge group suggests that an efficacious and ultimately emancipatory science of economics remains feasible. Moreover, once the constraints imposed by the implicit ontology implied by the unthinking commitment to formal modeling are lifted then progress can be made in developing powerful accounts of specific social existents such as the firm and the corporation by drawing on the general social ontology outlined by the group.

References

Deakin, S. 2017. Tony Lawson's theory of the corporation: Towards a social ontology of law. *Cambridge Journal of Economics* 41 (5): 1505–1523.

Faulkner, P., and J. Runde. 2013. Technological objects, social positions and the transformational model of social activity. *Management Information Systems Quarterly* 37 (3): 803–818.

Faulkner, P., S. Pratten, and J. Runde. 2017. Cambridge social ontology: Clarification, development and deployment. *Cambridge Journal of Economics* 41 (5): 1265–1278.

Lawson, C. 2008. An ontology of technology. *Techne* 12 (1): 48–64.

———. 2017. *Technology and Isolation*. Cambridge: Cambridge University Press.

Lawson, T. 1997. *Economics and Reality*. London: Routledge.

———. 2001. Evaluating trust, competition and cooperation: A realist perspective. In *Competition, Trust and Cooperation*, ed. Y. Shinoya et al. Berlin: Springer.

———. 2003. *Reorienting Economics*. London: Routledge.

———. 2007. Gender and social change. In *The Future of Gender*, ed. J. Brown. Cambridge: Cambridge University Press.

———. 2012. Ontology and the study of social reality: Emergence, organisation, community, power, social relations, corporations, artefacts and money. *Cambridge Journal of Economics* 36 (2): 345–387.

———. 2014. A speeding up of the rate of social change? Technology, resistance, globalisation and the good society. In *Late Modernity: Trajectories towards Morphogenic society*, ed. M.S. Archer. New York: Springer.

———. 2015a. What is an institution. In *Social Ontology and Modern Economics*, ed. S. Pratten. London: Routledge.

———. 2015b. *Essays on the Nature and State of Modern Economics*. London: Routledge.

———. 2015c. The nature of the firm and the peculiarities of the corporation. *Cambridge Journal of Economics* 39 (1): 1–32.

———. 2015d. The modern corporation: The site of a mechanism (of global social change) that is out of control. In *Social Morphogenesis: Generative mechanisms transforming late modernity*, ed. M. Archer. Berlin: Springer.

———. 2016a. Social positioning and the nature of money. *Cambridge Journal of Economics* 40 (4): 961–996.
———. 2016b. Collective practices and norms. In *Morphogenesis and Normativity*, ed. M.S. Archer. New York: Springer.
———. 2016c. Comparing conceptions of social ontology: Emergent social entities and/or institutional facts? *Journal for the Theory of Social Behaviour* 46 (4): 359–399.
Pratten, S. 2013. Community, rights and the self: Comparing critical realism, George Herbert Mead and Beth Singer. *Revue de Philosophie Economique* 14 (1): 73–103.
———. 2015. *Social Ontology and Modern Economics*. London: Routledge.
———. 2016. Coase on the nature and assessment of social institutions. In *The Elgar Companion to Ronald Coase*, ed. C. Menard and E. Bertrand. Cheltenham: Edward Elgar.
———. 2017. Trust and the social positioning process. *Cambridge Journal of Economics* 41 (5): 1419–1436.

Chapter 4
Central Fallacies of Modern Economics

Tony Lawson

Abstract Although it is widely recognised that the modern discipline of economics is short on explanatory successes, there is little sign that ongoing critical assessments of the situation are leading to any improvements. The reason for this lack of progress, I argue, is a prevalence of a set of fallacies maintained very often by mainstream practitioners and heterodox critics alike. These tend to take the form of presuppositions that underpin more explicit beliefs and accepted practices. Mostly they remain implicit and largely unnoticed. When noticed, they are regarded as so obviously correct they are rarely critically examined. Here I do examine them both explicitly and critically.

4.1 Introduction

> When you are criticising the philosophy of an epoch do not chiefly direct your attention to those intellectual positions which its exponents feel it necessary explicitly to defend. There will be some fundamental assumptions which adherents of all the variant systems within the epoch unconsciously presuppose. Such assumptions appear so obvious that people do not know what they are assuming because no other way of putting things has ever occurred to them. With these assumptions a certain limited number of types of philosophic systems are possible, and this group of systems constitutes the philosophy of the epoch. Alfred North Whitehead (1926, 61).

Intellectually speaking, the modern discipline of economics is in some disarray, short on explanatory successes, largely detached from its subject-matter, and seemingly without clear objectives or sense of direction. This is especially true of the hugely dominant mainstream project whose protagonists insist that mathematical modelling is the only proper or serious or 'scientific' way of doing economics (see Lawson 1997, 2003, 2015a).

T. Lawson (✉)
University of Cambridge, Cambridge, UK
e-mail: tony.lawson@econ.cam.ac.uk

© Springer International Publishing AG, part of Springer Nature 2018

P. Róna, L. Zsolnai (eds.), *Economic Objects and the Objects of Economics*, Virtues and Economics 3, https://doi.org/10.1007/978-3-319-94529-3_4

The latter project in particular has attracted much criticism of course, especially in recent years. Nevertheless it has doggedly survived almost unscathed and remains as dominant as ever. Why or how has this been possible given the continuing explanatory failures and indeed persistently unrealistic formulations?

One significant reason this mainstream project is able to persist in its unhappy state and yet simultaneously remain hugely dominant and influential, is the continuous impact of a set of fallacies, many and perhaps most of which are accepted at least as much by heterodox critics of the mainstream as by the latter's own protagonists.

These myths and fallacies usually take the form of presuppositions that underpin more explicit beliefs and accepted practices, and as such they tend very often to go unnoticed, or, if noticed, they are rarely critically examined. It is because the (often unwitting) acceptance of such fallacies is so harmful for the discipline that here I focus on formulating the more fundamental ones explicitly and in somewhat stark terms, placing them upfront, and indicating in an equally stark manner why indeed they are erroneous.

There are likely various reasons that fallacies of the sort that I have in mind do not receive greater critical examination. The most significant explanation though is simply that modern economists, not least those who set themselves up as media commentators, policy advisors/analysts and the like, mostly (there are important exceptions) reveal themselves to be unwilling to do the philosophical legwork necessary to get to the nub of the issues involved.

Most of these individuals unhesitatingly presume that the recourse of criticising substantive claims (typically modelling assumptions) of others and thereafter substituting (equally questionable) alternatives of their own is always sufficient and proper procedure (the sort of error that I am here seeking to dispel). To the extent that a few of the individuals in question do reveal some awareness of somewhat philosophically oriented critiques, the resort typically is to avoid the effort of engaging by instead displaying overtly dismissive postures, suggesting for example that their formulators know no economics, have hidden agendas, cannot do the mathematics, are 'economic flat-earthers', merely hide behind terms ending in 'ism' and 'ology', and so forth. The inevitable consequence is that discussions of the state of the modern discipline remain largely superficial, criticism is mostly misdirected and overly tame, and supposed/proposed alternative approaches or projects (some of which receive significant financial backing) end up, in the main, being essentially more of the same.

It is thus with the aim of counterbalancing tendencies of the noted sort that in the current paper I take the opportunity to focus solely, and in an explicit and sustained fashion, on those fallacies that, in my assessment, individually or collectively, serve most to obstruct serious attempts to transform modern academic economics into a more relevant, open-minded, serious and pluralistic discipline. I proceed, as I say, by way of formulating a number of them in simple and stark terms and indicating briefly why I take the presupposition in question to be erroneous. I also provide references to textual sources where the relevant argument is developed at greater length.

The fallacies I have in mind run as follows:

4.2 Twenty Fallacies of Modern Economics

Concerning the nature and problems of the discipline:

(1) *The widely observed crisis of the modern economics discipline turns on problems that originate at the level of economic theory and/or policy*

It does not. The basic problems mostly originate at the level of methodology, and in particular with the current emphasis on methods of mathematical modelling. The latter emphasis is an error given the lack of match of the methods in question to the conditions in which they are applied. So long as the critical focus remains only, or even mainly or centrally, at the level of substantive economic theory and/or policy matters, then no amount of alternative text books, popular monographs, introductory pocketbooks, journal and magazine articles, newspaper columns, blogs, *even student protests, petitions, and 'reclaiming economics' campaigns and events*, new institutes and/or centres, alternative programmes, conferences, workshops, plenary speeches, videos, comic strips, or whatever, are going to get at the nub of the problems and so have the wherewithal to help make economics a sufficiently relevant discipline. It is the methods and the manner of their usage that are the basic problem.

The point is simply that all methods are appropriate under some conditions but not others. Hammers and pens have their uses. But if the task at hand is, say, to mow the lawn neither a hammer nor a pen is likely to be up to the job. Similarly the sorts of mathematical methods economists insist upon have their uses. But social analysis is not one of them. This is because the methods in questions, to be successful, require closed systems, i.e. those in which correlations occur, where the guaranteeing of the latter closures require worlds of isolated atoms. By an atom I just mean a factor that always acts with the same, separate and independent effect, whatever the context. By isolated I merely mean that the economic atoms act in conditions in which there is nothing to interfere with their workings and so to prevent their effects from being deducible/predictable.

It is easy enough to demonstrate that social reality is not like this. Rather social reality is generally open, with everything (from social structures to embodied personalities) in process, being transformed through human practice (thus undermining atomism), with all aspects constituted in relation to (and not merely linked to and certainly not organised independently of) each other (thus undermining any requirement of isolationism). For lengthy discussions see for example Lawson 1997, 2003, 2015a.

(2) *The failings of modern economics emerged only with the recent economic crisis.*

This is simply false. Economics has been in an intellectually sorry state for the last 50 years or so. The output of the discipline has long been explanatorily a failure, plagued with unrealistic assumptions, and produced by those with no real idea where the project is going. This has been the case indeed ever since the significant take-up of methods of mathematical modelling in economics.

This sorry situation, moreover, has often been noted even by certain prominent mainstream spokespeople, not least *Nobel Memorial Prize Winners in Economics*, at least when adopting a reflective mode (in presidential speeches and such like). Thus for example back in 1982 Wassily Leontief was concerned that:

> Page after page of professional economic journals are filled with mathematical formulas leading the reader from sets of more or less plausible but entirely arbitrary assumptions to precisely stated but irrelevant theoretical conclusions.....Year after year economic theorists continue to produce scores of mathematical models and to explore in great detail their formal properties; and the econometricians fit algebraic functions of all possible shapes to essentially the same sets of data without being able to advance, in any perceptible way, a systematic understanding of the structure and the operations of a real economic system (Leontief 1982, 104).

Staying in the last century we find Milton Friedman lamenting the fact that:

> economics has become increasingly an arcane branch of mathematics rather than dealing with real economic problems (Friedman 1999, 137).

And Ronald Coase noting that;

> Existing economics is a theoretical system which floats in the air and which bears little relation to what happens in the real world (Coase 1999, 2).

An outline of the history of the failure of the discipline, as well as additional prominent acknowledgements of the latter's dire state can be found in, amongst other places, Lawson (2003 Chaps. 1 and 10) and Lawson (2015a, Chaps. 1, 5, 6, 7 and 11).

(3) *It is a failing of modern economics not to have predicted the timing of the recent crisis, (and given that so few did so, those who were successful should be lauded).*

Not really. Much has been made of the failure of economists to predict the fact, and precise timing, of the recent crisis. Even the British Queen has gotten in on the act in criticising economists for this perceived failure. Unfortunately all such criticism has only encouraged economists in the idea that event prediction is the legitimate goal to pursue, i.e., that the only serious economics is one (a form of mathematical modelling) that devotes resources to the aim of achieving this.

This is all quite misguided. Social reality including the future is open, so that successful event prediction is typically not much more creditable than winning a lottery. This is not to deny that we can understand many of the various tendencies in play at any time, not least those that are unsustainable. In the latter case we all know that something somewhere must 'give' in some way sooner or later (and the insight that capitalism will experience repeated crises, along with analyses as to why, has been provided by Marx and others a while ago). But when and how a specific manifestation happens is usually highly contingent.

Of course, as with all forms of betting it is very often the case with economic forecasting that, at each point in time, all possible outcomes are covered by the totality of forecasts made. Thus, at any given point in time there are usually some modellers that can claim (with at least some subset of their projections) to have got it 'right' (in the sense that official figures or 'measurements' fall with the assumed-to-be-appropriate bounds of error attached to these forecasts) whatever the actual outcome (although these 'official figures' are frequently revised with time, albeit perhaps after reputations have been established). But like any other gambles, those forecasts interpreted as successful are usually not followed by similar successes the next time around.

In any case, successful event prediction, if mostly infeasible in the social realm, is not only unnecessary but mostly unwanted anyway. This is so because we can still successfully identify explanatory social causes of phenomena, and the impossibility of systematic event prediction is consistent with our ability to use explanatory insights achieved to knowledgeably transform social reality to make the world a sounder place. In other words, it is better that we can knowledgeably (and preferably with wisdom) make our own history than are confined to merely watching it unfold, if albeit in a predictable manner (see Lawson 2015a, Chaps. 6 and especially 7).

(4) *The economics taught in modern universities is driven by right-wing or neo-liberal ideology.*

It is not. It is driven by a methodological ideology: that mathematical modelling is the only sound way to do economics. Throughout the modern academy the latter is widely accepted uncritically as 'common sense'. Most academic economists, in my experience, have little idea what neo-liberalism even means; nor do they care (see Lawson 2015a, Chaps. 3 and especially 7).

(5) *Dismissing the substantive theory and/or policy contributions of opponents with the label 'neoclassical' is helpful in pinpointing the problems of the discipline.*

It is not. Rarely is the term defined. And far from pinpointing or facilitating an understanding of any fundamental problem of the discipline it almost always detracts from doing so, through giving the impression that the problem is self-evidentially essentially a matter of poorly, but freely constructed, theory or policy -- that which is dismissed as neoclassical. This merely encourages lazy dismissals in the place of sustained explanatory analysis and critique (see Lawson 2013a, 2015a, Chap. 4). Not much better is the strategy of supposing that any approach labelled heterodox or Post Keynesian or institutional or whatever, especially if it is a form of mathematical modelling, is necessarily any better *just because* it is so labelled.

To transform the discipline in a constructive way it is necessary first to identify its problems; this is not achieved merely by signalling opposition or support for particular contributions through the use of unexplained labels (see Lawson 2013a or Lawson 2015a, Chap. 4 where it is argued that there actually is no neoclassical economic theory or policy).

(6) *Whatever may be the source of the discipline's problems we can always make progress by highlighting and chipping away at the lack of realisticness of prominent substantive assumptions (such as the familiar claims made about rationality, preferences, beliefs and states of the economy).*

I doubt it. To render mathematical modelling exercises tractable (to guarantee that embedded hypotheses are consistent with event correlations) economists must, as already noted, turn the 'agents' of their analyses into metaphorical atoms and situate them in isolated systems quite unlike those in which we actually live (see Lawson, 1997, 2003, 2015a). This endeavour (given the actual nature of human beings and the social world) necessitates the making of assumptions that are inevitably mostly unrealistic. It is specifically absurd formulations of rationality and the like that serve the purpose of 'atomisation', of reducing the 'subjects' of their analyses to preprogrammed (human) atoms.

Significantly however, although the cause of this lack of realisticness (i.e., the very use of inappropriate methods) goes unappreciated, *the fact of this lack of realisticness* of assumptions is usually acknowledged by everyone, including those who continually employ them, and frequently with regret; very often the assumptions made are interpreted as temporary devices that are expected to be improved upon in due course (1).

Certainly, few if any modellers or mainstream 'theorists' have defended their rationality assumptions as realistic, and when pushed, do often substitute other (fixed) behavioural assumptions or formulations (e.g., alternative specifications of rationality; or those of limited rationality; or those even of irrationality [as for example proposed by neuro-economists and the like]) that serve similarly to 'atomise' the subjects of the analysis.

In the same fashion, modern accounts of human beliefs (e.g. rational expectations) and economic states (e.g. equilibrium) work to facilitate model tractability or achieve various model consistency properties, and are rarely held by their formulators to be realistic. The former assumption merely renders the model in the head of posited agents consistent with the model in which the agents are situated by the economic modeller (see Lawson 1995). The latter equilibrium notion is just a solution concept of a system of (always unrealistic and irrelevant) equations Lawson 2015a, Chap. 8). There is rarely a pretence that any of this is expresses reality (see Lawson 2015a, especially Chaps. 4, 8 and 9).

In consequence, it is more or less futile for critics to think that inroads can be made by noting specific cases of assumptions that are unrealistic. Any lack of realisticness is rarely news; very often, as I say, it is regretted.

The insight that does seem like news, that does appear to go unrecognised by most, is that it is the emphasis on methods of mathematical modelling that is responsible for this persistent lack of realisticness, and that in an open complex social reality the production of unrealistic formulations is not a temporary contingent state but inevitable. Much better then to focus the critique on the modelling emphasis *per se* (see Lawson, 2015a, especially 4, 8 and 9).

(7) *The project of seeking to mathematise the economics discipline is a relatively modern one, and its dominance has been achieved through the project's significant explanatory successes.*

Not at all. It is a project that has been underway for over 200 years (see Lawson, 2003 Chap. 10). And the current dominance of the mathematical modelling endeavour within the academic discipline owes nothing to explanatory successes, something to the manner in which mathematics was reinterpreted by mathematicians themselves in the early years of the last century, and much to politics, especially following World War II.

For much of the history of attempts to mathematise economics, the aim was to do so in a manner that was analogous to the model of mechanics that dominated the non-social sciences. Economics was notably poor in this endeavour. However, in the early years of the twentieth century, a transformation took place in the manner that mathematics itself was interpreted. Basically the classical reductionist programme in physics fell into disarray as developments in relativity theory and especially quantum theory caused the image of nature as continuous to be re-examined in particular. A result was that the role of infinitesimal calculus, which had previously been regarded as having almost ubiquitous relevance within physics, came to be re-examined even within that domain.

One result, facilitated by the contributions of David Hilbert (1990 [1971]) in particular, was that mathematics became interpreted as (no longer an attempt to apply the physics model, and specifically the mechanics metaphor, but rather as) a discipline properly concerned with providing a pool of frameworks for *possible realities*. Mathematics was now viewed *not* as the language of (non-social) nature, abstracted from the study of the latter, but as a practice concerned with formulating systems comprising sets of axioms and their deductive consequences, with these systems in effect taking on a life of their own. The task of finding applications was henceforth regarded as being of secondary importance at best, and not of immediate concern.

This emergence of the axiomatic method removed many of the most severe constraints facing those who would seek to get mathematical practices accepted amongst the collective practices of the discipline of economics. For the time being, at least, researchers involved with mathematical projects in economics could postpone the day of interpreting their preferred axioms and assumptions. At this point mathematical modelling practices were included and able to compete amongst the accepted (collective) research practices of the discipline without the need to be formulated in ways that were analogous to the methods of mechanics, etc. Mathematical models indeed were created supposedly without carrying any necessary interpretation (2).

This development in the interpretation of mathematics moreover fed into an arguably even more significant environmental shift, this time concerning the set of political practices. The latter include both the onset of World War II and most especially the post-war emergence of McCarthyite witch-hunts in the US the face of the ongoing Cold War.

Initially, following the changes in the non-social sciences, the project of mathematising economics received its greatest stimulus in Austria and Germany, where the new physics, a revised conception of the role of mathematics and a specific emphasis upon axiomatic mathematics, had originated and came to flourish. In particular, it was here that von Neumann, Wald, Morgenstern and other mathematicians made their initial contributions. Approaches such as those of Wald and von Neumann were different in kind. But they were later reconciled in the US, where many of the early contributors emigrated under the Nazi threat.

So with attempts to mathematise economics freed up from the burden of having to fit with reality, it is significant that many of the practitioners of this approach ended up being located in the US in the post war period. This is so for two reasons especially. First it turned out that the US had the resources to dominate the post-world war II international academic scene in economics (as indeed it has done so in so many other disciplines). Second, this emphasis on an economics without any necessary interpretation proved to be an attractive proposition in the context of the McCarthyite witch-hunts in the face of the Cold War.

At that time the nature of the output of economics faculties - traditionally a form of political economy, a field that attracted those who sought a more humane system than capitalism - became a particularly sensitive matter. In such a context the project of mathematising economics proved to be especially attractive in that it carried scientific pretensions whilst being significantly devoid of any necessary empirical content.

The group most feared and mistrusted by the McCarthyites were the intellectuals (Reinert 2000). The mathematising project with its detached technicist emphasis, often to the exclusion of almost any critical or reflexive orientation, was clearly extremely attractive to those caught up in the situation. For the option of supporting an economics taking the form of such a project was extremely convenient not just to insecure or fearful university administrators but also to the funding agencies of US social scientific research (who were especially important in this period - see for example, Coats 1992; Goodwin 1998; Yonay 1998). Clearly by allocating funding to the mathematising project there was little risk to these bodies of being accused of supporting those who wanted to transform the economic system; for by everyone's account the mathematising project had no obvious bearing on social reality.

It was in this way that the mathematising project (first allowed to compete openly following the change in the way mathematics itself was interpreted) came to dominate the discipline, initially in the US, and eventually worldwide, despite having never been found to be explanatorily very successful (on all this see especially Lawson 2003, Chap. 10; also Lawson 2015a, Chaps. 1, 4, 5, 7 and 11).

Concerning theoretical/philosophical issues:

(8) *Economics can and should avoid philosophy*

It is this belief I am suggesting that most serves to prevent the discipline from identifying the obstacles that lie in the path of an emancipated economics. Most immediately, the widespread opposition to methodology serves to prevent criticism of the mainstream mathematical modelling emphasis, as well the development of alternatives.

But perhaps even more fundamentally, the neglect of philosophy in the form of ontology is an even greater obstacle to progress, allowing fallacies of the sort here being criticised to prevail almost unchallenged. Ontology is the study of the nature of being, as well as our implicit preconceptions about the latter. As such it is an inherently critical activity casting a spotlight on the presuppositions of all our activities.

This activity though is not limited to grounding critiques of standard practices. Like all forms of philosophy, ontology plays a systematic ground clearing role for science. As a result scientists, whatever their domain of study, need repeatedly to engage in it at significant moments in the advancing of causal, and indeed all other forms, of knowledge. Many physicists for example concern themselves with investigating the basic material of reality when they inquire into the *nature* of quantum fields, 'dark matter', particles and waves, mass, curved space-time, quantum gravity, black holes, etc., all issues in ontology.

Economics too has its more basic concerns. These include such matters as social relations, collective practices, social positions, community, capitalism, money, corporations, technology, gender, rights, obligations, human nature, care, trust, crises, economy, and so forth. Yet most economists, if inevitably occasionally referencing such categories, do rarely investigate their nature. However, it is impossible to provide much insight without at least some understanding of the nature of both social being in general and also the specific social phenomena being 'theorised'. These issues, all concerns of social ontological analysis, are easily shown to constitute part of the subject-matter of any would-be serious social science (see Lawson 2012a, b; 2015a, b, c, e, f; also see Archer 2013, 2014, 2015; Fullbrook 2009; Lawson et al. 2007; Mäki 2001; Pratten 2015; or Searle 1995, 2006, 2010).

(9) *The division between modern mainstream economics and heterodox alternatives rests fundamentally on competing substantive and policy claims.*

It does not. The division rests ultimately on very different ontological presuppositions (preconceptions, often implicit and unexamined, about the nature of social reality) combined with the fact that heterodox, but not mainstream, economists embrace pluralistic stances at the level of method.

Whilst the implicit ontological presuppositions of the reliance on mathematical methods of the sort on which economists often insist are closed systems/worlds of isolated atoms, it is easy enough to demonstrate that a far more explanatorily successful ontological conception finds social reality to be characterised by a prevalence of open systems marked by (internal-) relations, process, emergent totalities, meaning, value, etc. A concern with each of these latter features is found to characterise the various heterodox groups. Thus Post Keynesians concern themselves with openness is stressing the fact of fundamental uncertainty; feminist economics focus on relationality in their concern with care, gender, oppression, discrimination, emancipation, and so on; (old/original) institutionalists concern themselves with process (and forms of stability) in their analyses of social evolution, institutions, habit, technology, etc.; Marxian economists are concerned with that emergent internally-related totality-in-motion that is capitalism; and so on.

It is true that although this alternative ontology does not provide conditions for the generalised successful application of methods of mathematical modelling, many heterodox economists seemingly fail to recognise this and consequently allocate far more resources to experimenting with mathematical methods than appears reasonable. However heterodox mathematical modellers remain distinct from their mainstream counterparts in both not insisting that all economists employs such methods, and accepting the value of insights obtained by other means (on all this see Lawson, 2015a, especially Chap. 3, but also Chaps. 4 and 9).

(10) *To criticise/oppose the current mathematical modelling emphasis is to adopt an anti-mathematics stance.*

It is not. It is simply to point out that various tools (methods of mathematical modelling) are being used, and more or less exclusively so, in conditions (social reality) where these tools are generally inappropriate and more useful alternatives are available.

One of the many features that I find striking in the scene captured in Gustave Courbet's 1849 painting *The Stonebreakers,* depicted on the cover of my 2015 book, *The Nature and State of Modern Economics*, is the seeming incongruity between the wielding of relatively light-weight hand-held hammers and the task at hand – which is to break up rocks and stones in the process of creating a road through a hill or mountain. The use of mathematical modelling methods to address economic phenomena turns out to be at least as incongruous, and perhaps significantly more so. For when Courbet was painting, the hammers were conceivably used in the absence or unavailability of alternative tools that were more appropriate to the allotted task; or at least this was likely the situation facing the older man and young boy depicted (individuals who in themselves seem equally ill-suited to the sorts of tasks undertaken). However, it is not clear that mathematical models, interpreted as tools for providing insight to social reality, have ever been found to be useful to this allotted task; and there have always been better more productive/fruitful alternatives readily available (as well as people skilled in their application).

The point is that if close study of the practices of the modern economics academy reveals a situation that is every bit as marked by monotony and (albeit in a different sense) impoverishment as Courbet's scene (and its portrayal is similarly irritating to the powers that be) the rendering of this situation in a realistic if somewhat stark manner is *not* to oppose (or be 'anti') the tasks or the tools. Rather it is to criticise the generalised matching of each to the other along with the societal causal conditions that underpin the continuing insistence/conviction that the identified tasks be addressed or tackled only in the depicted ways (see Lawson 2003, 2009a, 2015a).

(11) *To criticise/oppose the current mathematical modelling emphasis is to adopt an anti-science stance.*

It is not. Mathematics is not essential (or inessential) to science; science involves using tools that are appropriate to the given task. A science of economics is perfectly feasible, and the current emphasis on mathematical modelling in economics serves, given the nature of social reality, mostly to prevent that potential from being realised (see Lawson 2015a, Chaps. 1 and 9).

(12) *To criticise/oppose the current mathematical modelling emphasis is to adopt an anti-pluralist stance.*

It is not. Pluralism, I take it, is an orientation of support for variety at all levels, as well as of tolerance and respect for, and willingness to listen to, and to engage with, others. To criticise the current emphasis on mathematics is not to argue for keeping the approach of mathematical modelling out of the toolbox, or to refuse to engage its users. Rather, it is to resist the dogma that *only* mathematical modelling methods should be, and be unquestioningly, utilised (and utilised however unrealistic the assumptions and explanatorily unsuccessful the whole endeavour; and despite the availability of more appropriate alternatives). It is, in other words, to resist one particular denial of pluralism, the version that currently dominates the discipline of economics (see Lawson, 2015a, Chaps. 6 and especially 10).

(13) *The mathematical models of modern economics can be shown to generate insights about aspects of the real world, once or if these models are appropriately, albeit super-cautiously, interpreted.*

This claim, where it is not totally banal, or a mere expression of hope and/or faith, is almost always based on a failure to recognise that, in most cases certainly, any insight attributed to the modelling endeavour was never actually a result of the latter, but rather achieved prior to model construction and incorporated into the modelling process. Mathematical models in the main are incapable of providing new insights about social reality; rather insights about the latter are tagged onto (bits of) the former where feasible in an attempt to make the models constructed appear in some way credible (see especially Lawson, 2009a).

(14) *Methods of mathematical modelling are, even if unnecessary, used in a neutral fashion, serving as just another language or heuristic device*

They are not used in a neutral fashion. They are tools. And like all tools they are appropriate for some tasks and conditions and not others. In certain contexts tools used inappropriately can be positively harmful. This has been (and is usually) the case with the application of mathematical methods in economics. It has forced the discipline into irrelevancy at best, whilst diverting resources away from potentially insightful alternative projects and applications. The claim that the mathematical methods adopted by economists are, or might conceivably be, employed as useful heuristic devices, serves, in the main, merely as an apology for this unhappy affair (see Lawson 1997, 2003 and especially Lawson 2009a).

(15) *Thought-to-be false assumptions and questionable modelling methods are justified and so useable if/where they generate agreeable conclusions, or anyway conclusions held to be true.*

This is incorrect, though seemingly widely believed even, or perhaps especially, amongst heterodox economists critical of the mainstream. That is, heterodox economists frequently suppose that although their modelling assumptions are (necessarily) false, their models are better (than those of their opponents) because the conclusions generated are held to be true. It may be true that ‘all polar bears are white'. But if this apparent truth is deductively generated from the assumptions that 'all polar bears eat snow' and 'all snow-eaters are white', we have added nothing to our understanding of polar bears, snow or whiteness; and nor have we provided explanatory support for the proposition that 'all polar bears are white'. All deductive exercises that are so based on known absurd fictions, and this inevitably includes almost all mathematical modelling exercises in modern economics, are just as pointless. Certainly they add little to our understanding of social reality (see Lawson, 2015a, Chaps. 5 and 6).

Concerning Proposals for Constructively Transforming the Discipline

(16) *The solution to making modern economics more relevant lies either in revising certain assumptions of mathematical models, or in a turn to more complex (in particular non-linear) forms of mathematical modelling, perhaps in the form of simulation analysis.*

It does not. This fallacy is based on a failure to see that the worldview presupposed by a reliance on these revised methods and forms is just as unrealistic (and indeed in essence is much the same) as that presupposed by more traditional ways of mathematical modelling. There is little reason to suppose that any of the novel modelling assumptions, modelling forms, model applications, or model estimation techniques currently on offer are of much use in the endeavour of rendering the discipline more relevant. The problem in all cases remains a mismatch of method and the conditions of application. A continuing inability to recognise, or reluctance to accept, this fact of the situation explains the failure of the more recent 'alternative' projects interpreted as critical and 'new'. The *Institute for New Economic Thinking* (INET) sponsored by George Soros is a particular example. Although INET no doubt sponsors a few projects that do avoid the noted problems, in the main, and despite Soros' own best intentions (see Lawson, 2015a, Chap. 9), the enterprise mostly fails to address the discipline's more fundamental problems, and indeed risks constituting an enormous waste of resources and opportunity (see Lawson 2015a, Chaps. 5, 6 and 7; see also Morgan 2015).

(17) *If conditions of experimental control do not hold in the social realm, then not only is science impossible, but all methods must be inadequate and not just mathematical modelling techniques. Thus in seeking to improve the discipline we might as well stick with the current emphasis on mathematical modelling.*

This is another view that is both pervasive and wrong. A social science can fruitfully concern it itself not only with much needed social ontological elaboration but also and especially with identifying unknown causes of significant phenomena, amongst other things.

Especially important here are dialectical methods such as contrast explanation. The latter proceeds by seeking to explain *not* some bear outcome Y but a (surprising) contrast along the lines of "Y rather than X", in conditions (a contrast space) where background knowledge leads us to expect all outcomes to be similar (to be X if this emerges as the typical and so expected outcome). For example why are these cows (with symptoms we now associate with mad cow disease) behaving differently to the rest? Why is the crop yield twice the average at the end of the field? Why do I feel so much worse today than usual? Why are house prices increasing far faster in one area of a city than the rest? And so on.

By asking such a question in a situation where we find a surprising contrast, we are effectively standardising for all the factors common to the cases being contrasted in the hope of identifying the causal factor responsible for the difference. This is the prion in the case of mad cow disease; perhaps a passing river in the case of crop yield; perhaps over-drinking of alcohol in the case of the individual not feeling too good; perhaps a train station in the case of an area of faster-rising houses prices, marked by a recent opening of a fast rail link to London; etc. (on all this see especially Lawson, Lawson 2009b, Lawson 2012a, b). Why is contrast explanation a dialectical method? Simply because the explanatory exercise starts (stage 2) with an apparent contradiction to (i.e., a surprising contrast in the context of) some sort of already achieved (stage 1) background understanding (all cows behave in similar sorts of way; the crop seed allocated, soil composition, and weather etc., is the same throughout a give field; etc.), where the eventual the outcome attained (stage 3) is an advance in understanding (achieved by resolving or accounting for the apparent contradiction). (On all this see Lawson, Lawson 2009b, Lawson 2012a, b; or Jamie Morgan 2013).

(18) *Economics, including any transformed discipline, can and should avoid matters of ethics/morality.*

Both parts of this claim are false. Ethics and moral argument are unavoidable, so it is better and indeed vital to address moral and ethical concerns in an explicit, systematic and sustained fashion (see Lawson, 2013b, 2015d). Even to contend that economists should avoid making judgements for purposes of practice/policy is to engage in ethical/moral argumentation. And to argue that almost all resources in the discipline be allocated to the practice of mathematical modelling is morally highly questionable (and indeed indefensible – see Lawson 2015d). Ethics, and specifically an ethics grounded in ontological analyses of such matters as human nature, care, the nature of social organisation, and the possibilities for flourishing (of human and other living beings), is essential for any suitably transformed more pluralistic, emancipated, economics (see especially Lawson 2015d).

(19) *Economics appropriately conceived is basically descriptive common sense, and this must be the basis of a transformed economics.*

This is not so. This mistaken view is unhelpfully widely promoted by various heterodox economists especially. Although mainstream and heterodox economists disagree on the value of descriptive common sense, the two are united in presuming that the latter is the only real alternative to methods of mathematical modelling (the latter being a methodological common sense to the mainstream – see Lawson 2015a, Chap. 6). This shared presumption is simply wrong. In fact, we all need seriously to raise our game and move way beyond common sense in all its forms; and in this both causal analysis and explicit, systematic and sustained projects in social ontology are likely essential. That is instead of simple and lazy common sense (and a naïve rush to forming simplistic policy recommendations) we need to return to pursuing economics in the manner prosecuted by the likes of Smith, Ricardo, Marx, Marshall, Veblen, Keynes, Schumpeter, Hayek and others, who dedicated themselves to explicit systematic and sustained programmes of uncovering the workings of the social system in which we actually live (see Lawson 2015a; and also see chapters in Stephen Pratten 2015).

(20) *The improving of economic teaching inevitably requires a good deal of prolonged, collective, formal discussion and debate over issues of substantive theory and policy*

This may be so, but I doubt it. Such an assessment seems to be underpinned by the idea that current academic economists, in constructing syllabuses, either conform to the uncaring atoms of their own theories or are ideologically biased at the level of theory and/or policy. I find neither to be typically the case. Rather, most modern economists are ideologically blinkered primarily by the idea that economics must be mathematical if it is to make a contribution. Even the lack of pluralism that so characterises the modern economics academy, stems mostly from a belief that to allow other methods into the toolbox will lead to a dumbing down of the discipline and a waste of resources.

Once the methodological error underpinning all this is revealed, or rather fully recognised, the potential is there in principle for the skills and energy of all the various participants in the economics academy to be harnessed to help fashion a more relevant discipline. Of course, new skills will likely need to be acquired by many. And serious research requires sustained effort and critical reflection. But we always must start from 'here'. In a process of successfully emancipating the discipline, the emphasis in the beginning will doubtless be as much upon supplementing, as upon replacing, existing courses, albeit likely involving the conversion of various currently compulsory courses into options (and also a change in styles of teaching [with the latter likely being rendered more interactive]). However, I do not anticipate that, with methodological blinkers removed, the task of providing relevant sets of courses in economics would be significantly more difficult (given time and sustained critical activity) than it is in any other open-minded, confident and successful discipline, where balancing acts regarding content taught are always to be performed (all disciplines must cope with issues of change in subject matter, competing interpretations and interests, and limited resources).

The pedagogical balancing act in an emancipated discipline of economics would presumably always be one of combining insights of the discipline regarded (albeit provisionally) as the more 'foundational' with any ongoing (possibly widely contested) advances regarded as contemporarily exciting and/or novel, taking into account local research and teaching expertise, skills and interests as well as the concerns of students; a transformed balancing act to that currently in play, but no less (or more) an inevitable balancing act.

The current problems, as I say, do not, in my assessment, derive in the main from an incapacity to care, or the sway of political ideology, or even an inability to find solutions. Rather, to repeat one last time, they stem from a pervasive and uncritical, indeed blinkered, belief in, and insistence upon, a simple and understandable, if ultimately mistaken, methodological dictum: that mathematical modelling is essential to any serious contribution to economics. Once that fallacy is transcended, the discipline can hopefully recommence its journey as a fruitful form of academic enquiry.

4.3 Conclusion

It is the declared goal of many observers both inside and outside the academy to improve the state of the current discipline of economics. My objective with the current paper is merely to point out that all such effort is almost worthless, certainly hopeless, if carried out without first emancipating ourselves from the noted methodological blinkers that render a (continuing) irrelevant economics inevitable. On a more positive note, it seems reasonable to suppose that once such blinkers are indeed removed, the obstacles that remain to achieving an emancipated and relevant discipline will be of the challenging and interesting sort that confront any academic endeavour. Of course there will remain various (political and other) differences no doubt, as in any science. But still achieving relevance and purposeful debate (if not agreement) about the state of the world has to be a very major step forward.

Notes (1) Of course, the regret is not always admitted. Some contributors even seek to turn a limitation into a virtue. Lucas for example writes: "To observe that economics is based on a superficial view of individual and social behaviour does not seem to me to be much of an insight. I think it is exactly this superficiality that gives economics much of the power that it has: its ability to predict human behaviour without knowing very much about the makeup and lives of the people whose behaviour we are trying to understand" (Lucas 1986, "Adaptive Behaviour and Economic Theory", *Journal of Business*, volume 49, No. 4, p. 425).

Mathematical models can certainly generate predictions (just as the latter can be plucked from the air). The feature they manifestly lack is the ability to achieve (in an open system such as the social world in which we actually live) is systematically to predict *at all accurately or successfully*. The point, though, is that even those as misguided as Lucas recognise the unrealisticness or "superficiality" of the assumptions made.

Lucas's focus is models employing the rational expectations hypothesis of course. Frank Hahn understands the possibilities for mathematical modelling better, explicitly rejecting the goal of successful prediction and observing instead that "What a rational expectations theory provides is an understanding of an imagined economy which satisfies the assumption." (Hahn 1985: 11, 12). Indeed, Hahn is often dismissive (to say the least) of Lucas and other who think such models might inform policy analysis: "When policy conclusions are drawn from such models, it is time to reach for one's gun" (Hahn 1982, p. 29). Rather, Hahn recognises that the goal can only be to see where unrealistic assumptions lead, or to enable particular outcomes to be generated: "When a mathematical economist assumes that there is a good economy lasting two periods, or that agents are infinitely lived [...] everyone can see that we are not dealing with any actual economy. The assumptions are there to enable certain results to emerge and not because they are to be taken descriptively" (Hahn 1994, p. 246).

Hahn, if clear that mathematical modelling is merely about exploring where unrealistic assumptions lead, never actually tells us why this activity is interesting or useful. The point here, though, is simply that it is evident that little difference/progress is going to be made by critics merely indicating that various assumptions regularly employed by mathematical modellers are not realistic. This is unlikely to be received as news, or as insightful, by anyone.

(2) Consider for example Debreu's eventual (Debreu 1959) axiomatic treatment of (the existence and uniqueness) of general equilibrium, a contribution that gained its author the Nobel Memorial Prize in economic science. Even today the language and symbolism of Debreu's *Theory of Value* is found in many axiomatic papers. Debreu's contribution rests for its legitimacy precisely on the claim that axioms are not in need of any interpretation. As Debreu expresses these matters himself: "*Allegiance to rigor dictates the axiomatic form of the analysis where the theory, in the strict sense, is logically entirely disconnected from its interpretations.* In order to bring out fully this disconnectedness, all definitions, all the hypotheses, and the main results of the theory, in the strict sense, are distinguished by italics; moreover, the transition from the informal discussion of interpretations to the formal construction of the theory is often marked by one of the expressions: "in the language of the theory," "for the sake of the theory," "formally." Such a dichotomy reveals all the assumptions and the logical structure of the analysis. It also makes possible immediate extensions of that analysis without modification of the theory by simple reinterpretation of the concepts; [...]" (Debreu 1959, p. x, emphasis added).

Acknowledgements The paper is a reprint of Tony Lawson: „Central Fallacies of Modern Economics" Schmollers Jahrbuch: Vol. 135, Wirtschaftswissenschaft(en) – Quo vadis?, pp. 189–208. Reprinted with the kind permission of the publisher.

References

Archer, S.M., ed. 2013. *Social Morphogenesis*. New York: Springer.

———., ed. 2014. *Late modernity: Trajectories towards the morphogenetic society*. New York: Springer.

———., ed. 2015. *Social morphogenesis: Generative mechanisms transforming the social order*. New York: Springer.

Coase, R. 1999. Interview with Ronald Coase. *Newsletter of the International Society for New Institutional Economics* 2 (1).

Coats, A.W.B. 1992. *On the History of Economic Thought: British and American Economic Essays*. Vol. I. London/New York: Routledge.

Debreu, G. 1959. *The theory of value*. New York: Wiley.

Friedman, M. 1999. Conversation with Milton Friedman. In *Conversations with leading economists: Interpreting modern macroeconomics*, ed. B. Snowdon and H. Vane, 124–144. Cheltenham: Edward Elgar.
Fullbrook, E., ed. 2009. *Ontology and economics*. London/New York: Routledge.
Goodwin, C. D. 1998. The patrons of economics in a time of transformation, In Morgan, M. S. and Rutherford, M. (Eds.). *From Interwar pluralism to postwar neoclassicism, Annual Supplement to Volume 30, History of political economy*, Duke/London: Duke University Press.
Hahn, F.H. 1982. *Money and inflation*. Oxford: Basil Blackwell.
Hahn, F. 1985. *"In praise of economic theory", the 1984 Jevons Memorial Fund lecture*. London: University College.
———. 1994. An Intellectual Retrospect. *Banca Nazionale del Lavoro Quarterly Review*: 245–258.
Hilbert, D. 1990 [1971]. Foundations of geometry *[Grundlagen der Geometrie]*, translated by Leo Unger from the 10th German edition (2nd English ed.), La Salle, IL: Open Court Publishing.
Lawson, C., J. Latsis, and N. Martins, eds. 2007. *Contributions to Social Ontology*. London/New York: Routledge.
Lawson, T. 1995. The Lucas critique: A generalisation', *Cambridge Journal of Economics*, vol. 19, No. 2, April, pp. 257–276.
———. 1997. *Economics and reality*. London/New York: Routledge.
———. 2003. *Reorienting economics*. London/New York: Routledge.
———. 2009a. On the nature and roles of formalism in economics. In *Ontology and economics*, ed. E. Fullbrook, 189–231. London/New York: Routledge.
———. 2009b. Applied economics, contrast explanation and asymmetric information. *Cambridge Journal of Economics* 33 (3): 405–419.
———. 2012a. Ontology and the study of social reality: Emergence, organisation, community, power, social relations, corporations, artefacts and money. *Cambridge Journal of Economics* 36 (2): 345–385.
———. 2012b. Emergence and social causation. In *Powers and capacities in philosophy*, ed. J. Greco and R. Groff, 285–307. London/New York: Routledge.
———. 2013a. What is this 'school' called neoclassical economics? *Cambridge Journal of Economics* 37 (5): 947–983.
———. 2013b. Ethical naturalism and forms of relativism. *Society* 50 (6): 570–575.
———. 2015a. *The nature and state of modern economics*. London/New York: Routledge.
———. 2015b. The nature of the firm, and peculiarities of the corporation. *Cambridge Journal of Economics* 39 (1): 1–32.
———. 2015c. The modern corporation: The site of a mechanism (of global social change) that is out-of-control? In *Social morphogenesis*, ed. M. Archer, 205–230. New York: Springer.
———. 2015d. Critical ethical naturalism: An orientation to ethics. In *Social ontology and modern economics*, ed. S. Pratten, 359–387. London/New York: Routledge.
———. 2015e. A conception of social ontology. In *Social ontology and modern economics*, ed. S. Pratten, 19–52. London/New York: Routledge.
———. 2015f. The nature of gender. In *Social ontology and modern economics*, ed. S. Pratten, 485–506. London/New York: Routledge.
Leontief, W. 1982. *Letter in Science* 217: 104–107.
Lucas, R.E. 1986. Adaptive behaviour economic theory. *Journal of Business* 59 (4).
Mäki, U., ed. 2001. *The economic world view: Studies in the ontology of economics*. Cambridge: Cambridge University Press.
Morgan, J. 2013. Forward looking contrast explanation, illustrated using the great moderation. *Cambridge Journal of Economics* 37 (4): 737–758.
———. 2015. Pluralism, Heterodoxy, and the prospects for a new economics curriculum: Assessing the potential of INET, What's the use of economics, and the CORE project, Mimeo: The Management Committee of the Association for Heterodox economics.
Pratten, S., ed. 2015. *Social ontology and modern economics*. London/New York: Routledge.

Reinert, E.S. 2000. The Austrians and the other canon. In *The history of evolutionary economics*, ed. J. Backhaus. Aldershot: Edward Elgar.
Searle, J.R. 1995. *The construction of social reality*. London: Penguin Books.
———. 2006. Social ontology: Some basic principles. *Anthropological Theory* 6 (1): 12–29.
———. 2010. *Making the social world: The structure of human civilization*. Oxford: Oxford University Press.
Whitehead, A.N. 1926. *Science and the modern world*. Cambridge: Cambridge University Press.
Yonay, Y.P. 1998. *The struggle over the soul of economics: Institutionalist and neoclassical economists in America between the wars*. Princeton: Princeton University Press.

Part III
Ontology of Modern Economics

Chapter 5
Social Scientific Naturalism Revisited

Daniel M. Hausman

Abstract The paper reconsiders social scientific naturalism, the view that despite obvious differences in their subject matter, the social sciences belong to the same species of cognitive inquiry as the natural sciences. Among other limits, the paper explores social scientific naturalism only with respect to economics. The social sciences are not homogeneous, and although many of the things I shall say apply to psychology, political science, sociology, and anthropology as well as to economics, I do not have the space to consider the many varieties of social inquiries. After a historical introduction in Section 5.1, Sections 5.2 and 5.3 clarify what is being asked when one raises the question of whether economics is relevantly similar to the natural sciences and why the question is worth asking. Section 5.4 then considers arguments to the effect that the subject matter of economics – namely human beings and their behavior – cannot be the subject of scientific inquiry. Sections 5.5 and 5.6 focus on an important peculiarity of economics and the other social sciences, namely, that it employs concepts that are continuous with those of daily life. This fact turns out to have considerable ramifications. Section 5.7 concludes.

The organizers of the symposium on Ontology and Economics and of the larger project "to investigate the reconstruction of economics as a moral science" are, among other things, calling for a reconsideration of social scientific naturalism – that is, the view that despite obvious differences in their subject matter, the social sciences belong to the same species of cognitive inquiry as the natural sciences. Whether social scientific naturalism is a tenable view is the question that got me started in philosophy of social science more than 40 years ago. Although I cannot claim to have devoted 40 years to this question, I have tossed it around many times in my mind, and I hope in this essay to clarify and narrow the question, if not to resolve it once and for all. Among other limits, I shall explore social scientific naturalism only with respect to economics. The social sciences are not homogeneous, and although many of the things I shall say apply to psychology, political science,

D. M. Hausman (✉)
University of Wisconsin-Madison, Madison, USA
e-mail: dhausman@wisc.edu

© Springer International Publishing AG, part of Springer Nature 2018
P. Róna, L. Zsolnai (eds.), *Economic Objects and the Objects of Economics*, Virtues and Economics 3, https://doi.org/10.1007/978-3-319-94529-3_5

sociology, and anthropology as well as to economics, I do not have the space to consider the many varieties of social inquiries.

After a historical introduction in Section 5.1, Sections 5.2 and 5.3 clarify what is being asked when one raises the question of whether economics is relevantly similar to the natural sciences and why the question is worth asking. Section 5.4 then considers arguments to the effect that the subject matter of economics – namely human beings and their behavior – cannot be the subject of scientific inquiry. Sections 5.5 and 5.6 focus on an important peculiarity of economics and the other social sciences, namely, that it employs concepts that are continuous with those of daily life. This fact turns out to have considerable ramifications. Section 5.7 concludes.

5.1 Historical Introduction

As the conception of *science*, as distinct from philosophy, evolved in the eighteenth century, questions gradually emerged about whether inquiries into the principles governing social relations count as sciences. Given the striking success and prestige attached to physics, it is not surprising that those studying society sought to clothe themselves in the mantel of *science*. Thus one finds espousals of social scientific naturalism such as Condorcet's. In an address on the 12th of February, 1792 to the French academy, he looked to the social sciences to provide increasingly for our feelings, just as the natural sciences help to satisfy our basic material needs.

Crucial in the evolution both of social scientific naturalism and of social inquiry itself are discussions of what we now call economics. The notion of unintended consequences, so brilliantly delineated in the work of Hume (1752) and Smith (1776), identifies a gap between the intended goals of individual human agents and the outcomes agents actually bring about. This gap creates a space for systematic inquiry that does not reduce to journalism. By the mid-nineteenth century, when John Stuart Mill publishes his *System of Logic* (1843), social scientific naturalism is ascendant. Mill classifies economics and the social sciences unambiguously as the same sort of inquiries as the natural sciences. Like specific natural sciences, the social sciences have their peculiarities. For example, in economics, as in meteorology, there are a great many causal factors and the possibilities of experimentation and precise prediction are severely limited. But in Mill's view difficulties such as these are not peculiar to the social sciences and mark no distinction between the natural and the social sciences.

The second half of the nineteenth century saw an intellectual reaction to this assimilation of the social sciences to the natural sciences (Koslowski 1997). Many of the challenges to social scientific naturalism were focused on specific features of classical and then neoclassical economics – in particular, the use of extremely simplified and idealized models which facilitated the use of deductive reasoning and mathematical representation. There were also more general qualms grounded in the subjective character of economic concepts and generalizations and the influence of values in economics.

Qualms about the scientific status of the social sciences, including economics, have persisted, but they have had little impact on the conduct of economics itself. Although one finds complaints from leading economists that the field values mathematical ingenuity far more than empirical adequacy and that economics is biased by ideology and corrupted by special interests, the day-to-day conduct of economics continues to mimic the conduct of the natural sciences. Indeed, there has been a surge in laboratory and field experiments and in the use of techniques such as instrumental variables that permit more focused empirical testing than one finds in economics a generation ago.

Economists generally ignore or dismiss arguments that maintain that their aping of the natural sciences is misleading pretense, or that it plays a larger role in defending the status and salaries of economists than in illuminating the properties, causes or consequences of economic behavior. But the fact that economists are not disturbed by these criticisms is not a refutation of them. One needs to examine the arguments that purport to show that economics differs from the natural sciences in some relevant way in its goals, methods, concepts, ontology, or epistemic warrant. Amid all the smoke of criticism, where are the fires? What fuels them? If they can be stoked or extinguished, should they be?

5.2 Ambiguities: Are Social Inquiries "Like" the Natural Sciences?

The general question, "Is economics fundamentally different from the natural sciences?" is vague. "Different" in what regard? Do or should economic inquiries employ different concepts or have different ontological implications? Should they employ the same methods as the natural sciences? Do they have the same goals? Are their findings as credible?

In addition, distinguishing the different regards in which economics and other social inquiries may resemble or differ from the natural sciences leaves the questions about whether the social sciences differ from the natural sciences obscure, because the natural sciences differ among themselves with respect to their goals, concepts, ontology, methodology, and epistemic warrant. However, the debate over social scientific naturalism appears to presuppose that there is some set of features shared by all the natural sciences. Otherwise it is difficult to see how one can sensibly ask whether economics has the same features as the natural sciences.

One might reasonably doubt whether the features shared by *all* the natural sciences are substantial enough to make the general question of whether economics resembles the natural sciences worth asking. There are many differences among sciences such as physics, meteorology, medicine, and entomology. Laboratory experiments are crucial in some, impossible in others. Some aim at generalizations; others aims at knowledge of particular structures. Idealizations are ubiquitous in some and absent in others. Some permit precise predictions; others do not. Some theories of quantum phenomena even apparently countenance contradictions.

The list of common features is shorter: All the natural sciences are sensitive in some way or other to observational evidence. All seek a certain sort of objectivity, aiming either at truth or at an adequate (albeit possibly inexact) representation of what can be observed. All formulate theories or models. All make some use of mathematics, although the importance and kind of mathematics differ. If possession of these common features were sufficient to determine that an enterprise constitutes a science, then economics would count as a science – but so would astrology, homeopathy, or divination.

5.3 What Makes Social Scientific Naturalism Worth Probing?

A deeper question must be answered before proceeding further: Why should one care about social scientific naturalism? Why does it matter whether economics differs from the natural sciences in one regard or another? Suppose, for example, one concluded that economics differs from the natural sciences in its relevance to policies and to the material interests of individuals. So what? Why is this of any greater interest than the fact that microscopes are more valuable tools in biology than in astronomy? Knowing why differences between economics and the natural sciences may matter helps to identify which questions about the relations between economics and the natural sciences are important and what sorts of answers one is looking for.

I suggest that the most important question concerns *methodology*, in the sense of the most general principles governing knowledge acquisition, rather than in the sense of specific techniques such as Bayesian or classical statistics. If the general methods of acquiring knowledge that are employed in the natural sciences are not efficacious in economics, then the conduct of economics must be changed or reconceptualized, and its claims to have provided knowledge of its subject matter are cast into doubt. The other ways in which economics may differ from the natural sciences are important mainly because of their bearing on this central question concerning methodology. It is because differences in ontology, concepts, and goals may demand differences in the conduct of economics and in the credibility of its conclusions that these differences are worth exploring.

There are many efforts in the literature to establish explicit criteria distinguishing sciences from other sorts of cognitive enterprises. The most famous of these is Karl Popper's falsificationism (1959, 1962). It consists of two separable theses, although Popper does not himself separate them. The first is a claim about scientific theories: they are logically falsifiable – that is, they are logically inconsistent with some finite set of true or false statements about observations. This is a claim about a conceptual difference between sciences and other disciplines. Sciences employ falsifiable theories. Popper argues that this conceptual difference has important epistemic implications: there is no way to find out whether a logically consistent unfalsifiable theory is false.

Unfortunately, as Popper recognized, this claim (that scientific theories are logically falsifiable) is mistaken. To demand that scientific theories be logically falsifiable rules all probabilistic claims out of science and fails to take into account the fact that scientists must supplement significant scientific theories with various background theories or auxiliary assumptions in order to derive implications that can be falsified by observation.

Popper's second thesis, which helps to rescue a reinterpretation of the first, distinguishes scientific practices from other cognitive activities by a methodological commitment among scientists to submit theories to harsh testing and to refuse to make excuses for them when they fail the tests. What distinguishes sciences from other activities is methodological rather than conceptual. Even though scientific theories are *not* usually logically falsifiable, scientific norms call on practitioners to supplement theories with additional premises that, for the purposes of testing, they take to be unquestioned. From these supplemented theories, scientists derive implications that they believe to be especially vulnerable to falsification by means of experiment or observation. If after carrying out the tests, they find that the implications of the supplemented theory are false, scientists should not blame the various additional premises they employed to derive the implication they tested. They should instead decide that the theory is false and seek an alternative. Theories that pass harsh tests are worthy of further harsh testing.

Popper's characterization of scientific disciplines as governed by the "critical attitude" described above permits one to criticize the scientific pretensions of much of economics. Like scientists, economists derive implications of their theories (when supplemented with a variety of auxiliary assumptions concerning people's knowledge of the alternatives, their access to credit, how competitive the market is, and so forth) to make predictions concerning, for example, the effects of a change in taxation. But, unlike what Popper demands of scientists, economists do not reject or even seriously question their theories when these predictions are falsified. So goes the Popperian critique of the scientific claims of economics that one finds in works by Mark Blaug (1980) and Terence Hutchison (2000).

The problem with this critique is that it applies as readily to the natural sciences as to economics. Although some sciences permit much more precise predictions, whose failure gives rise to greater concern among the practitioners, no science is governed by Popperian norms. Theories are too valuable and too hard to generate to allow failed predictions to have such drastic consequences. Because the various auxiliary assumptions one must make in order to derive observable implications from a theory may be false, it is always open to question whether any observation or experimental result is relevant to one's assessment of a theory. In the case of economics, where the auxiliary assumptions are frequently extreme idealizations, it would be foolish to decide that failed predictions are always refutations of economic theory.

This line of thought is disquieting, because it seems to provide economic theories with immunity from disconfirmation. Economists can always blame an apparent refutation on some simplification, interference, or disturbing cause. If evidence can never disconfirm theories, why pay any attention to it? There's obviously much more to investigate concerning whether economists assess theories in the same way that natural scientists do. But Popper's views on the demarcation between science and other sorts of inquiries is of little help in that investigation.

5.4 Can the Behavior of Human Beings Be the Subject of a Science?

Human free will may suggest a metaphysical doubt about the possibility of a social science. Owing to free will, it would seem that regularities in the social sciences cannot have the character of natural laws. But free will clearly does not rule out the possibility of finding regularities in human behavior, which, in fact, abound. Even though not every entrepreneur is governed by the pursuit of larger net returns, to suppose that the pursuit of profits motivates entrepreneurs is a reliable generalization. But, the critic of social scientific naturalism continues, economists formulate such generalizations by imaginatively putting themselves in the situations of agents and asking what would be a "sensible" thing to do, rather than by gathering objective data. This generalization embodies a sort of maxim of rationality rather than a simple empirical generalization. I think the critic's claim is exaggerated. Observation and experiment play a crucial role, too. How else can economists learn about phenomena such as preference reversals (Lichtenstein and Slovic 1971)? Even if free will exists, which is far from obvious, and it places limits on lawful behavior in the social sciences, which is also questionable, regularities in human behavior remain, and social scientists are daily learning more about them via straightforward empirical methods.

A related but more specific concern accuses economics of reifying human nature. Having found that people in contemporary commercial societies are acquisitive, economists allegedly take greed to be a feature of human nature itself, ignoring the extent to which the behavioral regularities that are invoked to explain economic outcomes are themselves the product of economic circumstances. Since what people take to be human nature is not a pre-social given, but an evolving social product, economics goes wrong in imitating the natural sciences and treating the features of human nature (and economic rationality in particular) that it employs to explain and predict human behavior as if they were unchanging causes.

Presumably there have been economists who have thought that every human being is a *homo economicus*, but I doubt that such anthropological naïvety is common. What is more important is whether the fact that features of human character are evolving rather than constant undermines the scientific status of economics. Consider a generalization such as, (G) "People prefer larger bundles of commodities and services to smaller." This generalization is obviously not universally true, and its relevance to a social circumstance in which there are few tradable commodities or services is questionable. If explanation and prediction require exceptionless laws, then this generalization is of no explanatory or predictive worth. But such a conclusion is absurd. Indeed, in his influential book, *Making Things Happen*, James Woodward (2003) argues that explanation does not require laws. A critic of social scientific naturalism might more plausibly maintain that although generalizations such as G are valuable for both prediction and explanation, they cannot provide truly scientific explanations. Why not? More needs to be said.

A further complication that arises when one studies people is that the agents, whose reasons economists discuss, are capable of learning what economists say about them, which may in turn lead them to change their behavior. For example, when economists simplify and maintain that consumers, workers, and investors are all entirely self-interested, they may lead people to become more self-interested. In other contexts, the findings of economists may be self-defeating rather than self-supporting. Having learned about some of the cognitive failings identified by psychologists and behavioral economists, economic agents may take care to avoid these mistakes.

Although the existence of self-fulfilling and self-defeating results is an intriguing feature of social inquiries that has no counterpart in the natural sciences, it is hard to see how it leads to any conceptual, metaphysical, or epistemic divide between economics and the natural sciences. Theoretical descriptions and predictions need not be shared with the agents whose behavior they describe, and if they are shared, knowing them may have no effect on people's behavior. When the claims economists make have a causal influence on social actors, there is no reason why that influence cannot be incorporated into the theory's claims, which can then be tested like any other set of generalizations.

5.5 The "Verstehen" Requirement

The discussion so far has touched on four dimensions along which there may be differences between economics and the natural sciences. Are there ontological differences, conceptual differences, methodological differences, or epistemic differences? In addition, one might ask whether the goals of economic inquiry are, at the relevant level of abstraction, different from the goals of the natural sciences. Philosophers and sociologists have argued that in addition to or instead of the predictive and explanatory goals of the natural sciences, the social sciences should aim at providing us with *Verstehen (understanding)*. This issue receives its classic discussion in the work of Max Weber (1904) and followers such as Alfred Schutz (1932; Barber 2014).

This different or additional goal of the social sciences mandates conceptual and methodological differences. Weber and many others argue that the social sciences should provide understanding of social phenomena that is linked to the understanding possessed by social participants. Let us call this "the *Verstehen* requirement." Satisfying the *Verstehen* requirement enables social theorists to share the participant's grasp of the phenomena, although the theorist may build additional concepts such as Weber's ideal types, "on top of" the concepts that the participants employ. Following Weber, Schutz insists that understanding what others aim to do (rather than merely the causal antecedents of their actions) is essential both to the ability of social actors to sustain social reality and to the ability of social theorists to provide an accurate description and conceptualization of social reality. Because social reality is constructed out of everyday understanding, social theories must encompass those constructs in their own constructs, which must thus be linked to particular social structures rather than taking the form of universal generalizations.

In addition, Weber maintains that however interested theorists may be in regularities, people want to understand particular happenings in their details and individuality, rather than, as in the natural sciences, merely as instances of general regularities. Weber and Schutz argue that social theorists inevitably classify social phenomena in terms of various culturally significant or meaningful categories, and that explanations must be in these terms or they will not tell people what they want to know. This *Verstehen* requirement seems to introduce an element of subjectivity into the social sciences that is avoidable in the natural sciences. But, provided that social theorists conceptualize the phenomena in the proper meaningful terms, Weber has no objection to causal explanation.

Whether the *Verstehen* requirement calls for a change or reconceptualization of economic practice or undermines the credibility of economic conclusions is questionable. The concepts economists employ: expectations, preferences, choices, money, prices, quantities, trade, and so forth are all continuous with the concepts of everyday social experience and thus satisfy the *Verstehen* requirement. But, at the same time, as Weber envisioned, these concepts figure in generalizations that appear to be explanatory and predictive.

Some twentieth century philosophers who were influenced by Weber and by developments in the philosophy of language (especially the work of Wittgenstein) argued that the *Verstehen* requirement has more radical implications (i.e. Melden 1961; Von Wright 1971 and Winch 1958). In particular, they maintained that it is a mistake to assimilate regularities couched in the vocabulary of everyday social experience to regularities in the natural sciences. Regularities in the natural sciences derive from causal laws and specifications of initial conditions. They alleged that this is not true of most regularities in the social sciences, which, in contrast, result from rules, conventions, or institutions. They maintained that to explain these regularities requires identifying and understanding the conventions and rules that give rise to them rather than identifying causal laws. Grasping conventions and rules requires understanding meanings rather than experimentation. Winch (unlike Weber) thus maintains that understanding or interpretation supplants the empirical inquiry that characterizes the natural sciences.

Winch's view would demand a drastic reconceptualization of economics and a radical revision in its methodology. But his view is hard to accept. Surely there is a role for observation and experiment in assessing claims about the details of conventions, the extent to which people in a society conform to them, the functions conventions perform, or the sanctions attached to violating conventions. Regardless of whether the law of demand arises in part from conventions, one can still assess the causal generalization that, *ceteris paribus*, an increase in the price of some commodity *x* causes a decrease how much of *x* people will buy.

5.6 The Nitty Gritty Significance of the "Verstehen" Requirement

Although I do not think that Winch's critique of social scientific naturalism is tenable, the fact that economists employ everyday concepts such as belief and preference (in accordance with the *Verstehen* condition) has important implications. First, as macro-economists have come increasingly to recognize, beliefs, not just the realities about which people have beliefs, are crucial to understanding human behavior and guiding economic policy. For example, economists have argued that to spur the sluggish Japanese economy, Japanese investors and consumers must be convinced that the government and national bank will support a much higher rate of inflation. Unfortunately, the only way to convince Japanese consumers and investors of this is to carry out highly expansionist monetary policies for an extended period. (This is not to deny that in a good deal of economic modeling, beliefs and expectations play no role. Economists often simplify and assume that people have perfect information. But, as economists recognize, this is a large and often questionable idealization.)

Taking beliefs into account requires that economists distinguish between on the one hand facts and "objective" probabilities and, on the other hand, reality as it appears to economic agents, whose beliefs may be false and whose subjective probabilities may differ from the expectations of experts. Moreover, although some economists – defenders of revealed preference theory – pretend that preferences can be defined in terms of choices, once economists concede that beliefs are crucial to choices, they must recognize that preferences are subjective states that combine with beliefs to determine choices. One can infer preferences from choices (or vice versa) only given premises concerning beliefs, and one can infer beliefs or expectations from choices only given premises concerning preferences. Someone who inferred from the fact that Romeo drinks poison when he might have eloped with Juliet that Romeo was so repulsed by the prospect of life with Juliet that he drank poison to avoid eloping would be, to say the least, obtuse. One must understand Romeo's options when he drinks the poison *as they appear to him*.

Preferences and beliefs are thus "intentional" concepts (Quine 1960, Chap. 4, 6). Their objects need not be real. When a suspicious entrepreneur believes that a conspiracy among her competitors is causing her losses, there need not be any conspiracy. Intentional concepts have a different logic. From a true *non*intentional statement such as "The retail price of can of Pepsi is \$1," and the true second premise, "George paid the retail price for a can of Pepsi," one can infer "George paid \$1." But from the true *intentional* statement that "George believed that the retail price for a can of Pepsi is 50 cents" and the same second premise, it does not follow that "George paid 50 cents."

Alexander Rosenberg (1980) maintains that generalizations employing intentional terms such as belief and preference cannot be laws. I do not follow his argument, and in any case, Woodward (2003) has made a powerful case that what is needed to guide action and to enable explanations are generalizations that support counterfactuals, even if these are not universal or exceptionless. What is true of intentional terms is that they have a different logic, which reflects the fact that an agent's beliefs and preferences are concerned with a subjective object: the world as it appears to the agent. Clearly social life would be impossible if there were not a great deal of agreement among people's beliefs. Moreover, for those who are not insane, there are many points of contact between their subjective reality and the way things actually are. But people's beliefs and desires can be directed toward non-existent objects and states of affairs, and mistaken beliefs have a huge role in economic life. Without them, markets for equities and bonds would be very boring places.

The fact that economics operates with concepts that satisfy the *Verstehen* requirement, has an additional implication which further separates economics from the natural sciences. Explanations in economics in terms of preferences and expectations, like everyday explanations in terms of beliefs and desires, state the *reasons* why agents act. Reasons, unlike causes in the natural sciences, are subject to *assessment*. They may be criticized as confused, not well considered, insufficient, imprudent, or immoral and, of course, they can be defended from these criticisms and praised as noble, wise, generous, or astute. These issues concerning the justification of reasons, which have no counterpart in the natural sciences, are not just additional considerations, independent of causal questions about the role of reasons in explaining behavior. Agents are usually aware of their reasons, even if they sometimes deceive themselves or are compelled by unconscious psychological drives. If they cannot defend the reasons in favor of carrying out some action, then they are unlikely to carry out that action and, if they do nevertheless perform the action, the reasons in favor of the action will not explain why.

So long as economics relies on the concepts of preferences and expectations, its explanations of behavior will cite agents' reasons. Thus, it is inevitable that economics is built on a theory of rational choice. For the possession of reasons presupposes rational capacities. Only a being capable of rationality can judge whether a reason counts in favor or against carrying out some action or judge a reason's force or quality.

The *Verstehen* requirement also suggests that economics will be more heavily influenced by values than are the natural sciences and that it will be more closely tied to policy. Because economics is built around a theory of rationality and is capable of issuing assessments of the rationality of beliefs, preferences and choices, which assessments are in turn accessible to economic actors, economics has an unavoidable role in guiding individual conduct and social policy. Economics is, minimally, attached to the value of rationality itself. In addition, because theories have social consequences, theorizing is subject to moral scrutiny. Not only must the conduct of economics respect the moral constraints on all scientific practices, which require truth telling, courtesy, avoiding personal attacks on those with whom one disagrees, and so forth, but economists must also be cognizant of the possibility that their theories do harm, by (for example) revealing to unscrupulous agents possibilities for exploiting the rational foibles of others. Furthermore, as Weber emphasized, the choice of questions economists ask, will inevitably be influenced by their importance to the interests of individuals and social groups. The fact that values influence both the conduct of economics and the questions economists investigate does not, of course, imply that the answers economists give to the questions they ask will be biased by their personal ideals or by their political commitments.

That economics is relevant to policy is uncontroversial, but the natural sciences and engineering are relevant to policy, too. Some social scientific naturalists maintain that economics serves policy in just the same way that the natural sciences guide policies – that is, by helping policy makers choose means that will achieve their ends. Such a practical role for scientific knowledge is familiar and straightforward. Agents have some goal that they want to accomplish, and the scientist provides information on how to achieve that goal. One way in which economics influences policy is surely as a source of information about the consequences of alternative policy choices.

The *Verstehen* requirement suggests that there are deeper connections between economics, policy, and values. First, as already mentioned, the concerns of private interests and policy makers influence which questions social theorists ask. In addition, they constrain the range of possible solutions that are seriously considered. The advice economists offered concerning the proper response to the near collapse of the global financial system in 2008 was surely constrained by the power of large financial interests. The influence of what Marx (1867) called "the Furies of private interest" can sometimes be crude: economists, like physicists or philosophers can be "bought," and there is much greater demand for pliant economists than for corrupt metaphysicians. Customs and social values also make economists unwilling to defend positions that appear to be irresponsible or unreasonable. For example, although the logic of cost-benefit analysis implies that the economic cost of a person's death depends on the person's expected income, it is customary in cost-benefit analysis to assign the same cost to deaths at the same age, whether the deceased was rich or poor.

Though it is hard to deny that customary values and ideological forces have influenced many economists, it is difficult to quantify how large an effect they have had. Overt bias is less common than more subtle influences. Since policy makers rarely turn to economists with precisely formulated goals, economists may help determine the goals. The major economists of the past two centuries have also been social philosophers who have found in economic theory support for their social ideals. Economists such as Alfred Marshall, Milton Friedman, and Kenneth Arrow wrote their economics in the service of their political philosophy. In saying this, I am not maintaining that their economics is faulty or biased. Nor am I claiming that their political philosophy was not in turn influenced by their work in economics. The point is rather that, dealing with matters of importance to people's everyday lives, serious economic inquiries are inevitably intertwined with the investigator's ideals of social life.

I conclude that the *Verstehen* requirement casts a long subjective and evaluative shadow across economics. Because economics aims to describe, predict, and explain a social reality that is in large part constituted by the beliefs and preferences of individuals, it is forced to theorize at least partly in those terms. It is thus forced to take account of the subjective perspectives of agents and to cope with the intentionality of belief and preference. Moreover, the explanations economists offer of individual choices may be assessed not only for the accuracy of their causal attributions but also for the reasonableness of the beliefs, preferences, and choices that they attribute to individuals.

5.7 Conclusions

The discussion above establishes eight features of economics that distinguish it from most or all the natural sciences:

1. The subject matter of economics is human behaviour, its presuppositions and outcomes.
2. The generalizations of economics are not universal. They concern features of human beings and the social world that change over time.
3. The objects of economics can be influenced by what economists say about them. This complicates the assessment of economic claims and imposes moral constraints on what economists announce.
4. The explanatory variables of economics are continuous with the concepts employed by economic agents.
5. The central explanatory concepts of economics – beliefs and preferences – are intentional and not directly observable. Beliefs can be inferred from behavior only if observers make assumptions about preferences, and preferences can be inferred from behavior only if observers make assumptions about beliefs.
6. Explanations in economics of choice behavior cite the reasons for actions, and they thus assess beliefs, preference, and choices as well as explain them.
7. Economics is built around a theory of rationality, and it is thus in part a normative enterprise, at least with respect to rationality.
8. Economics is so important to people's interests that its findings may be distorted by social norms, corruption, and ideology.

These differences are clearly significant. Economics is in many ways unlike any of the natural sciences. But these differences do not undermine the ambition to learn about economies by formulating models, including mathematical models, and testing applications to specific explanatory and predictive problems. There are many ways in which economics (and the other social sciences) differ from the natural sciences. Economics is a distinctive enterprise, very unlike physics. But economics nevertheless puts forth generalizations that are not completely insulated from empirical refutation; and despite its causal complexity and the powerful influence of economic interests, it makes progress in identifying, explaining, and predicting economic phenomena.

References

Barber, M. 2014. Alfred Schutz. In *Stanford Encyclopedia of Philosophy* https://plato.stanford.edu/entries/schutz/.
Blaug, M. 1980. *The methodology of economics: Or, how economists explain*. Cambridge: Cambridge University Press.
Hume, D. 1752. *Political Discourses*. Rpt. New York: HardPress Publishing, 2012.
Hutchison, T. 2000. *On the Methodology of Economics and the Formalist Revolution*. Cheltenham: Edward Elgar.
Koslowski, P. 1997. The theory of ethical economy in the historical school. In *Contemporary theory*, ed. Wilhelm Roscher, Lorenz von Stein, Gustav Schmoller, and Wilhelm Dilthey. Berlin: Springer.
Lichtenstein, S., and P. Slovic. 1971. Reversals of preferences between bids and choices in gambling decisions. *Journal of Experimental Psychology* 8: 46–55.
Marx, K. 1867. *Capital*. Vol. 1, Trans. S. Moore and E. Aveling. New York: International Publishers, 1967.
Melden, A.I. 1961. *Free action*. London: Routledge and Kegan Paul.
Mill, J. S. 1843. *A System of Logic*. Rpt. London: Longmans, 1949.
Popper, K. 1959. *The logic of scientific discovery*. New York: Basic Books.
———. 1962. *Conjectures and refutations*. New York: Basic Books.
Quine, W. 1960. *Word and object*. Cambridge, MA: MIT Press.
Rosenberg, A. 1980. *Sociobiology and the preemption of social sciences*. Baltimore: Johns Hopkins University Press.
Schutz, A. 1932. *The phenomenology of the social world*. Trans. G. Walsh and F. Lehnert. Evanston: Northwestern University Press, 1967.
Smith, A. 1776. *An Inquiry into the Nature and Causes of the Wealth of Nations* Rpt. New York: Oxford university press.
Von Wright, G. 1971. *Explanation and understanding*. Ithaca: Cornell University Press.
Weber, M. 1904. Objectivity in social science and social policy in *The Methodology of the Social Sciences*, E. A. Shils and H. A. Finch (ed. and trans.), New York: Free Press, 1949.
Woodward, J. 2003. *Making Things Happen*. Oxford: Oxford University Press.

Chapter 6
Is Economics a Moral Science?

David Colander

Abstract Economics is a multifaceted inquiry into the social relationship among humans in society. This mutifacetedness creates a lack of communication between critics of economics (which includes a large proportion of philosophers and professors with a religious bent) and defenders of what, for lack of a better term, I will call mainstream economics. The problem is that when critics attack economics, their attack generally focuses on the one-dimensional vision of the social relationship of individuals that is conveyed by the principal texts rather than on a much more sophisticated vision of humans and of social relationships held by good economists. The author fully agrees that morals and science are intertwined in a Putnam sense—the questions one asks and the framework one chooses reflects moral choices. But the paper argues that that as a pragmatic way to move forward, that deeper intertwinement can be usefully separated from less deep, entanglements that are more easily understood and recognized.

Economics is a multifaceted inquiry into the social relationship among humans in society. This mutifacetedness creates a lack of communication between critics of economics (which includes a large proportion of philosophers and professors with a religious bent) and defenders of what, for lack of a better term, I will call mainstream economics. The problem is that when critics attack economics, their attack generally focuses on the one-dimensional vision of the social relationship of individuals that is conveyed by the principal texts rather than on a much more sophisticated vision of humans and of social relationships held by good economists.

While it may be true that some economists accept a "selfish MaxU homo-economicus" description of humankind, and also accept a positivist notion that good economics does not include moral judgments, most mainstream economists, myself included, accept neither of those propositions. We see humans as complex beings with social and moral goals that are culturally embedded into society and human decision making. We see the simpler textbook models, and the more compli-

D. Colander (✉)
Middlebury College, Middlebury, USA
e-mail: colander@middlebury.edu

© Springer International Publishing AG, part of Springer Nature 2018
P. Róna, L. Zsolnai (eds.), *Economic Objects and the Objects of Economics*, Virtues and Economics 3, https://doi.org/10.1007/978-3-319-94529-3_6

cated MaxU models upon which they are based, as useful logical structures—what Ariel Rubinstein (2012) calls fables. These fables are useful in thinking about policy, but they are not definitive. If these models are not tempered by educated common sense which integrates morality and sensibilities into their use, then they are highly misleading in providing policy guidance. We fully recognize that human interaction has a moral dimension, and that so too does economic policy. To the degree that economics has anything to say about policy, it must be a moral endeavor.

Whether that moral endeavor should be called a moral science is debatable, and has been debated for centuries. Debates that have been ongoing for centuries, without resolution, are either very deep debates indeed, or are debates in which ambiguity of definitions play an important role. I have little to add to the deep philosophical debate about economics' nature as a moral science. That debate is above my pay grade, and I leave it to philosophers such as Hilary Putnam (2002) and philosophically oriented economists such as Kenneth Boulding (1969). But I do have views about the definitional issues and the way economics is practiced. In my view, the debate about whether economics is a moral science has been hampered by a failure to communicate by the various participants in the debate. The problem is that the debaters are often using differing definitions of both science and economics. This leads them to often argue at cross purposes, with little actual communication. These definitional differences make the deep, possibly irresolvable, philosophical differences seem more important than they need be.

In this paper I have two goals. The first is to explain the role that I see definitional issues playing in the "Is economics a moral science?" debate. (My answer is that both sides are right. Economics is and is not a moral science depending on how one defines economics and science.) I argue that when definitional issues are removed, the debate is far less contentious, and the differences of view smaller. The second goal is to argue that critics of economics would be most effective if they direct their attack about economists' lack of moral foundations at economic pedagogy and applied economics practice, not at economic theory or the economics profession in general. Doing so avoids a debate about whether good economists know that economics is a moral endeavor, and focuses instead on whether economists convey that knowledge appropriately in their teaching of the principles of economics and their presentation of policy advice to policy makers. A strong actionable argument can be made that they do not.

6.1 Definitional Aspects of the Debate

For me, 95% of the "Is economics a moral science?" debate involves differing definitions of the terms "science" and "economics", not differing sensibilities about the moral nature of economic questions. Eliminate the definitional differences and there will be much more consensus among mainstream economists and critics of economics about the nature in which economic analysis is a moral endeavor.

The definition of economics I like best is attributed to Jacob Viner: "Economics is what economists do". For non-economists, this is not an especially helpful definition, but it does nicely capture the arbitrariness of any definition of economics. Since what economists do is varied, and changes over time, any definition will inevitably be incomplete and ambiguous. For this reason any specific definition should not be given too much weight in capturing what it is that economists do.

We can see this incompleteness and change in the evolution of textbook definitions. Adam Smith defined "political economy", the term he and other Classical economists, used as the term to designate their field of study, as "an inquiry into the nature and causes of the wealth of nations," Other Classical economist's definition had a different focus, but all emphasized that political economy involved the study of material welfare. How they studied human's pursuit of material welfare, and what aspect of it they studied, was something that changed over time as problems facing society changed.

This focus on material welfare is important for a consideration of economics as a moral endeavor. Classical writers focused on material welfare to emphasize that economists were studying just a portion of human welfare, not all welfare. They recognized that their formal economic MaxU homo economicus theory was based on assumptions that were unrealistic, and wanted to emphasize that the conclusions of their economic theory did not extend to policy. One of the first Classical economists to write explicitly about methodology, Nassau Senior, made this clear when he wrote:

> (An economist's) conclusions, whatever be their generality and their truth, do not authorize him in adding a single syllable of advice. That privilege belongs to the writer or statesman who has considered all the causes which may promote or impede the general welfare of those whom he addresses, not to the theorist who has considered only one, though among the most important of those causes. The business of a Political Economist is neither to recommend nor to dissuade, but to state general principles, which it is fatal to neglect, but neither advisable, nor perhaps practicable, to use as the sole, or even the principle, guides in the actual conduct of affairs (Senior 1836: 2–3).

6.2 Marshall's Definition and Methodology

In the late 1800s, as Classical economics evolved into Neoclassical economics, the definition of economics changed, as did the term that economists used to designate their broad field of study. Whereas Classical economics called their field "political economy", and often used the term "economics" as a designation of a theoretical sub branch of political economy that focused on a set of logical deductive models that may or may not be related to economic policy, neoclasicals began to use the term "economics" synonymously with political economy. The change was led by Alfred Marshal who started using the term "economics" to designate what Classical economists has designated as "political economy". Marshall made no distinction between the two branches. To emphasize the change, he called his textbook *Principles of Economics*, (Marshall 1890) rather than *Principles of Political*

Economy, as Classical economists such as David Ricardo (1817), John Stuart Mill (1848a) and Henry Sidgwick (1883) had called their texts.

Marshall also changed the definition of economics (which replaced political economy), writing that economics was "the study of mankind in the ordinary business of life; it examines that part of individual and social action which is most closely connected with the attainment and with the use of the material requisites of well-being." This definition kept the material welfare aspect of economics, but reduced the sense that an important branch of economics was a theoretical branch that focused on the reasoning process of what would later become known as homo economicus. It was that theoretical branch that Classical economists wanted to separate from the study of economic policy; Marshall wanted to go further and essentially eliminate it. For Marshall, economics and political economy were seen as one, and the economics that mattered was focused on the real world and policy, not on theory with its abstract models populated with homo-economicuses. Abstract complex theoretical models based on homo economicus had no place in Marshall's blend of theory and policy. For Marshall economics was a science without a formal theory.

6.3 Robbins' Definition and Methodology

The profession did not follow Marshall and in the 1930s economics changed. Marshall's downplaying of MaxU theory, and emphasis on the ordinary business of life, lost favor, and was replaced with Walras' general equilibrium theory built on MaxU assumptions. It did however follow Marshall in replacing "political economy" with "economics" as the description of the broad field. This was problematic because abstract Max U theory was once again on the rise, making the distinction between economics and political economy relevant once again. The integration of multivariate calculus into economist's theory extended its reach and importance in reasoning, cementing marginal analysis into all economists' thinking, and making the homo economicus model central to the teaching of economics.

As this change was happening there were considerable debates about the scope and method of economics. In this debate Lionel Robbins offered a different definition of economics, and it is a variant of that definition that one is most likely to see in the texts today. That definition moved away from the material welfare focus emphasized by Classical economists and also away from the real-world policy focus of Marshall. Robbins (1932) defined economics as "the science which studies human behavior as a relationship between given ends and scarce means which have alternative uses." This was a return to the Classical emphasis on economics being directly involved with developing a deductive logical theory, which could not be directly applied to policy. But it also involved a broadening of its focus of theory away from material welfare, and toward a broader choice theory relating to all aspects of life, not just to material aspects. Thus, it abandoned the material welfare limitation that previous economists had used in the definition to remind them that economic theory left out important elements of reality and could not be directly applied to policy issues. Robbins believed that the theory of rational choice that economists were developing at the time related to all choice, not just to material goods, and was simply the analysis of constrained optimization.

It is Robbins' definition of economics, which is closely connected to Robbins' argument that the science of economics should not include value judgments, that most critics refer to when they criticize economics for excluding morals.[1] The problem is that this criticism has more to do with definitions than with substantive differences. Specifically, Robbins did not see himself as defining the broad economic endeavor. Rather he saw himself as defining a sub branch of the economic endeavor—which at various times has been called economic science or positive economics.[2] To make this clear he entitled his book, *The Nature and Scope of Economic Science*, not The Nature and Scope of Economics. It is in this sub branch where homo economicus resides; he does not reside in the real world. Since the analysis of rational choice theory related to all rational choice, not just to choices about material goods, the material welfare limitation was dropped. The purpose of that limitation was to remind economists of the limitations of applying economic analysis to real world problems. For Classical economists, the results of scientific analysis (the analysis of a world populated with homo economicuses) was not to be directly applied to policy, so the material welfare qualification was unnecessary. Consideration of that limitation belonged in discussions of the applied policy branch of economics, not in the discussion of economic science.

Robbins was not arguing that economics *should* study the allocation of scarce resources among alternative ends. His was a descriptive definition, not a delineating definition. If economics is what economists do, and what economic scientists were doing was exploring constrained optimization models that dealt with the allocation of scarce recourses among alternative ends, then that would work as a definition of economics. Of course, economic scientists did other things as well, so not much store should be put on that description definition. Robbins did not put much store on it. He writes: "Economics, we have suggested, is essentially the study of the disposal of scarce goods and services. This suggestion has not met with universal acceptance. But the difference between this and other definitions now current is not a very serious matter." (Robbins 1938; p. 344).

Once one understands Robbins' definitions, it is clear that his argument about morals and economics is an argument about *where* morals should enter in the broader field of economics, not *whether* they should enter in. He wanted economists to not follow Marshall in combining theory and policy into a single analysis, but instead to follow the Classical methodological approach of separating an abstract

[1] I discuss Robbins' work in detail in Colander (2009).

[2] Whether that sub branch is a science or not depends on the definition of science one is using, and early economists had many different definitions of science. They distinguished pure science from realistic science, and there were many shades of both. Alfred Marshall felt that all economics blended together and chose to call all of economics a science. Others distinguished a science which dealt primarily with logical truisms from broader economic inquiry which they chose to call an art and craft, not a science. I discuss these issues in Colander (1992).

scientific theory, which one did not attempt to apply directly to policy, from an applied policy branch that was explicitly concerned with integrating values and morals into the analysis. His methodology required maintaining a strict firewall between theory and policy analysis. So what Robbins wanted was for the science of economics to not include an analysis of morals since he felt that morals were best analyzed using a philosophical, not a scientific, methodology.

Critics of economics who argue that economics does not consider itself a moral science usually relate their argument to Robbins's definition, which tends to predominate in modern texts. They argue that economics sees itself as a narrow science, rather than as a broader moral science. While I agree that, in practice, some economists may think that they are excluding moral judgments from their analysis and from economics, Robbins was not one of them. Moreover, the broader mainstream, of which I see myself as a part, does not see this definition as excluding morals from the broader field of economics at all. All the issues have been explored in depth in work in social choice theory and welfare economics.[3] That is why when critics attack economics for not understanding the moral dimension of economics, economists such as myself feel that they are missing their mark.[4]

Robbins makes his views clear in his Eli Lecture in which he reflected on what he felt was the way his essay had been misinterpreted. He writes:

> My suggestion here, as in the Introduction to my Political Economy: Past and Present, is that its (political economy) use should be revived as now covering that part of our sphere of interest which essentially involves judgments of value. Political Economy, thus conceived, is quite unashamedly concerned with the assumptions of policy and the results flowing from them. I may say that this is not (repeat not) a recent habit of mine. In the Preface to my Economic Planning and International Order, published in 1937, I describe it as "essentially an essay in what may be called Political economy as distinct from Economics in the stricter sense of the word." It depends upon the technical apparatus of analytical Economics; but it applies this apparatus to the examination of schemes for the realization of aims whose formulation lies outside Economics; and it does not abstain from appeal to the probabilities of political practice when such an appeal has seemed relevant. It should be clear then that Political economy in this sense involves all the models of analysis and explicit or implicit judgments of value that are usually involved when economists discuss assessments of benefits and the reverse or recommendations for policy. (Robbins 1981: 8)

[3] Economist's resolution of these issues goes back to the 1930s when Gunnar Myrdal (1929 Swedish edition; 1932 German edition) pointed out the difficulty of moving from theory to policy. His work was only translated into English in 1952 but it was well known in its German translation. Abram Bergson (1938, 1954) also pointed out that economic analysis needed an externally determined social welfare function to have any possibility of moving from theory to policy implications. Later, Amartya Sen (1970) extended the analysis with social choice theory. It is that work, and the work of their students, to which I refer when I argue that the economics profession has accepted that morals must be integrated into policy analysis. The work of RDG Little (1950) and van Graff (1957) pointed out the difficulties with applying economist's theoretical model to practice, but that did not affect the belief of economists that they had theoretically solved the problem. The acceptance of the Walrasian general equilibrium model as the basic model for policy pushed all such considerations out of practical economic policy analysis, and the work is hardly discussed in the principles course.

[4] An example of what I mean can be found in the debate between Hillary Putnam and Partha DasGupta which I discuss in Su and Colander (2013).

6.4 The Moral Philosophical Foundation of Classical Economics

The moral philosophical foundation of Classical economics can be found in variations of utilitarianism which dates back to Jeremy Bentham (1780), John Stuart Mill (1859; 1863), and Henry Sidgwick (1883). For Classical economists utilitarianism was a moral philosophy, not a science. It provided guidance for political economy, not for the science of economics. It used a moral philosophical methodology, not a scientific methodology. Thus, truths in utilitarianism would be arrived at through what might be called argumentation for the sake of heaven—where debater's goal was to understand the arguments on all side so that they can jointly arrive at agreement on moral judgments to use in policy. Utilitarianism was a consequentialist philosophy—one judges a policy by its outcome, not by an outside code of morals. To make a utilitarian argument required an understanding of both economic science and economic practice—how the policy would play out in the real world. The two were often different.

Utilitarianism was also an egalitarian philosophy whose goal was "the greatest good for the greatest number". That goal made it explicit that the moral judgment should be from a social perspective that took into account all individuals, not just the social perspective of one group. To implement this approach required economists to use some variation of the impartial spectator mechanism that Adam Smith used in his *Theory of Moral Sentiments* (1759). It involved approaching problems from the position of an impartial spectator, and addressing one's policy argumentation in a way that would be able to gain the support of an impartial spectator. The process required careful philosophical exploration and extensive debate. Mill provides a nice description of the needed approach to reasoning that utilitarians had in mind.

He who knows only his own side of the case, knows little of that. His reasons may be good, and no one may have been able to refute them. But if he is equally unable to refute the reasons on the opposite side; if he does not so much as know what they are, he has no ground for preferring either opinion. The rational position for him would be suspension of judgment, and unless he contents himself with that, he is either led by authority, or adopts, like the generality of the world, the side to which he feels most inclination. Nor is it enough that he should hear the arguments of adversaries from his own teachers, presented as they state them, and accompanied by what they offer as refutations. That is not the way to do justice to the arguments, or bring them into real contact with his own mind. He must be able to hear them from persons who actually believe them; who defend them in earnest, and do their very utmost for them....So essential is this discipline to a real understanding of moral and human subjects, that if opponents of all important truths do not exist, it is indispensable to imagine them, and supply them with the strongest arguments which the most skilful devil's advocate can conjure up (Mill 1848b(1947)/1859: 35–36).

Variations of utilitarianism were the moral compass that Classical economists arrived at. Because it served as a moral compass, not a strict guide, utilitarianism did not require quantitatively measurable utility, since its role was to provide moral guidance in making the value judgments necessary in policy. The "greatest good for the greatest number" might have no mathematical formulation, and may even be a contradiction in terms, but it conveyed an understanding that was useful in thinking about policy.[5] It was useful in excluding certain arguments, not to distinguish between all arguments. If one is arguing for a policy because it benefits oneself, without concern for others, it does not meet utilitarianism criteria. Utilitarianism does not arrive at a single correct policy, but it does narrow the set of possible policies. Further resolution of differences in policy would have to be arrived at through argumentation for the sake of heaven.

The utilitarianism that Mill advocated did not accept that utility maximization of individual's perception of their utility should be the goal of policy. Mill distinguished between higher and lower pleasures, and noted Socrates' notion that it is better to be a human dissatisfied than a pig satisfied. In arriving at moral judgments all such issues had to be taken into account. For policy, judgments on such issues were important, but the judgment would not be arrived at through the use of the scientific methods, but instead through the use of philosophical methods.

Robbins did not oppose the Millian version of utilitarianism; he accepted that some moral foundation for policy was needed. He instead emphasized the need to keep morals out of science because he felt that utilitarianism had no scientific foundation. Utility wasn't measurable, and even it were measureable, it is not clear that maximizing utility was the goal of society.[6] The Robbins position was that since you cannot measure philosophical concept, such as utility, utilitarian issues had to be dealt with using a different moral philosophical methodology, not a scientific methodology.

The Classical liberal methodology that Robbins was advancing totally agrees that economics (the broad definition) is a moral endeavor, and could be called a moral science, if one chooses to define science broadly to include all types or structured inquiry. But it argues that a useful heuristic is to break up the field of economics into subcomponents--a narrow scientific inquiry which follows scientific methodology and is concerned with only finding the truth, and a wider applied policy branch that blends scientific and philosophical methodology. Within this heuristic, "economic *science*" is not a moral endeavor because, by design, science avoids moral issues because it doesn't have any good way of resolving differences of views on those issues.[7]

---

[5] Mill (1863) put it this way "The utility standard may be hard to apply, but it is better than having no standard."

[6] This philosophical concept of utility was not something that was designed to be measurable, as some economists later attempted to do with it. It was simply a constructed concept meant to provide moral guidance—to remind policy makers to look at policy from the position of an impartial spectator, not from the point of view of one side or another. That's why there were two different concepts of utility—one the philosophical concept, utility, utilitarian issues had to be dealt with using a different moral philosophical methodology, not a scientific methodology. (2007)

[7] This is a heuristic, not a definitive rule. It accepts that scientific analysis reflects some inseparable moral judgments, but it does the best it can to reduce these as much as it can.

The key point to note is that this resolution only works if there are other branches of economics that deal with moral issues. As discussed in J.N. Keynes (1891) classical economics had those other branches. Positive economic science was supplemented by normative economics in which the goals of economic policy were to be determined, and an art of economics where economists explored how best to achieve the goals, given understanding determined in the science. This third "art" branch of economics was by far the largest branch, and its methodology was a blend of philosophic methodology involving argumentation for the sake of heaven and scientific methodology involving a blend of deductive logical models, empirical findings of "facts" and testing of models.[8]

The economics profession did not follow Robbins. Instead, it attempted to blend the formal science and art of economics together to formally arrive at policy conclusions. Neoclassical Walrasian theorizing attempted to portray maximizing utility as the obvious goal of economic policy. This was quite outside the philosophical view of utilitarianism as an acceptable vague goal, but not an acceptable precise goal. The goal of utilitarianism was to provide a broad framework for philosophical debate about moral dimensions of economic issues. It was not something to be applied directly into economist's formal choice theoretic models or something to be used for direct policy advice. Economists' abandonment of the Classical method led to well-grounded criticisms of economics, and to the changes in formal welfare economic theory discussed above. In economist's welfare economics and social choice work the need for moral foundations for economic policy was formally recognized, but no guidance was provided as to how those moral foundations could be integrated into their real-world applied work or even into their teaching. Therein lies the problem.

6.5 What Is to Be Done?

The arguments I am making are not deep, and I argue, are understood and accepted by economists who have spent time studying the issue.[9] If accepted and acted upon, they could avoid much fruitless debate. I do not claim to have answers to the deeper questions about the entanglement of morals and science; I fully agree that morals and science are intertwined in a Putnam sense—the questions one asks and the framework one chooses reflects moral choices. But I argue that that as a pragmatic way to move forward, that deeper intertwinement can be usefully separated from less deep entanglements that are more easily understood and recognized. But to do

[8] For a discussion of how the art of economics was lost, see Colander (1992). See also Colander and Su (2015).

[9] I am not arguing that most economists have seriously thought about them about them. Welfare economics and methodology is hardly taught in graduate economic programs and thus most economists have little introduction to it. Like most individuals, economist do what they do, and spend little time thinking about broad questions of methodology.

that we economists need specific guidance from applied philosophers, not general or theoretical critiques of our failings.

My argument is that economists are willing to accept agreement about the second type entwinement, and they have built a separation of moral analysis from scientific analysis into their formal welfare analysis. Their social welfare function approach requires any policy pronouncement to specify an imposed social welfare function, embodying the moral judgments of the policy maker.[10] Theoretically, that acknowledgement goes a long way toward meeting the criticisms of critics who argue that economic policy is a moral endeavor.

Since sophisticated economists have accepted critic's arguments, and provided a theoretical answer, most mainstream economists feel justified in dismissing critics even though in practice, they do a lousy job of making the moral dimension of economic policy clear. It is because theoretically mainstream economists have recognized the problem that I argue that critics should avoid criticism of the profession per se. Critiques of the economics profession generally will draw critics into debates about economic theory that are, in my view, unwinnable by critics, and which will distract from what I believe is the primary concern of critics, which is not about deep economic theory, but about the way economics is practiced and the way it is taught. I agree with critics of economics, economic pedagogy and economic policy practices as they relate to integrating the moral dimensions of economics to students and policy makers are highly problematic. It is here where I suggest critics of economics direct their criticisms. The teaching of economics does not adequately convey to students that economic policy is a moral endeavor, and economic policy analysis does not make the moral dimension of economic policy clear.

Economists justify what they teach with the argument that simplification is necessary. While the need for pedagogical simplification is obvious, I believe that the specific pedagogical simplifications that are currently made by economics cannot be justified, and exist primarily because economists are taking the easy way out. The reality is that it is easier to teach semi-formal models populated by homo economicuses, than it is to teach the nuanced skill of relating abstract models into reality. The texts do not distinguish economic science from the boarder economic endeavor, and do not sufficiently warn students *not* to directly apply the theoretical models they

[10] One theoretical economist, Ariel Rubinstein, a well known game theorist, and author of a standard text in graduate microeconomics has been very clear about the need to distinguish the moral dimension to economic policy analysis. He writes: "I would like to start with what I believe every academic should do when appearing in public, especially when speaking about political and controversial issues—to clarify the extent to which he is incorporating his professional knowledge in his remarks, when he is expressing views with the authority supported by academic findings, and what part of his comments are nothing more than his personal thoughts and opinions….to the best of my understanding, economic theory has nothing to say about the heart of the issue under discussion here…Because as an economic theorist, I would like to state that economic theory is exploited in discussion about current economic issues, and I don't like it…, to put it mildly. Everything that I say here, even in an academic context (and I intentionally use the word "academic" since I do not think that the word "scientific" is appropriate for economics) is completely subjective, controversial and therefore perhaps describes me no less than it describes economic theory" (Rubinstein 2012: 13–14).

learn directly to policy questions. The texts do not give students sufficient practice in applying models, and the give essentially no introduction to the philosophical methodology that exists to guide economists in doing so. Philosophical utilitarianism is not presented to students; instead students are led to believe that somehow by avoiding measuring utility that they have avoided the problem of integrating morals into economic policy analysis. Instead of conveying to students the reality–that technical economics may provide some support for standard economic policy arguments, it along cannot answer policy questions without moral input–the books convey the sense that scientific positive economics can answer policy questions.[11]

6.6 Conclusion

Let me conclude by considering the role that I see philosophers playing in helping to resolve these issues. As I have emphasized in the paper I do not see that role as attacking the economics profession for not understanding the issues. Instead, I suggest a better route is attacking the pedagogical simplifications that the economics profession has made in its teaching of economics. That teaching concentrates far too much on learning technical models with far too little on the difficulties of applying those technical models to real world problems. It provides far too little discussion of why integrating moral judgments into the analysis is a necessary part of policy analysis. That teaching allows the simple textbook models to define the policy questions and policy solutions rather than to let common sense define them. The teaching of economics focuses far too little on teaching students that the goals of society must be collectively determined before any policy can be advocated, and on the philosophical methodology to arrive at those collective determinations.

Economists, on their own, are not going to change. They haven't been taught any of the philosophical methods that might provide guidance on these issues, so even if they wanted to, they are not prepared to teach students this aspect of applied policy economics. The principles of economics course is very close to the same course taught in 1950; it reflects a template that Paul Samuelson created. It is here where I see philosophers playing an important role. By directing their criticisms at economic pedagogy—analyzing the texts and pointing out where they go wrong, developing examples that show where moral arguments could nicely enter in, and possibly even creating a two or three class "philosophical economics module" that is designed for economists to use to integrate moral issues into the teaching of economics—they could make an enormous difference. Philosophy and religion

[11] One specific example of economic pedagogy failing is in its failure to deal with the possibility of endogenous tastes. If tastes are endogenous individual's utility functions are influenced by the operation of the system. In this case, one cannot take preferences as givens. Without some moral input about what tastes are preferable, one cannot say what policy is desirable. With semi-endogenous tastes the entire textbook economic policy foundation is undermined. This would seem to be an important policy issue. Yet, one sees no discussion of that possibility in the economic principles texts.

departments could even start giving a Masters and PhD in Philosophical Economics, with the goal of turning out professors who could have joint appointments in economics and religion and philosophy departments. These specialists could add the needed dimension. Economics is too important to leave the teaching of economics to economists alone.

References

Bentham, J. 1780/1907. *An Introduction to Principles of Morals and Legislation*. Oxford: Claredon Press.
Bergson, A. 1938. A reformulation of certain aspects of welfare economics. *The Quarterly Journal of Economics* 52 (2): 310–334.
———. 1954. On the concept of social welfare. *The Quarterly Journal of Economics* 68 (2): 233–252.
Boulding, K. 1969. Economics as a Moral Science. *The American Economic Review* 59 (1): 1–12.
Colander, D. 1992. The lost art of economics. *Journal of Economic Perspectives* 6 (3).
———. 2007. Edgeworth's hedonimeter and the quest to measure utility. *Journal of Economic Perspectives* 21 (2 Spring): 215–226.
———. 2009. What was it that Robbins was defining? *Journal of the History of Economic Thought* 31 (04): 437–448.
Colander, D., and H.-C. Su. 2015. Making sense of Economist's positive normative distinction. *Journal of Economic Methodology* 22 (2): *157–170.*
Graff, J. de V. 1957. *Theoretical Welfare Economics*. Cambridge: Press Syndicate of the University of Cambridge.
Keynes, J.N. 1891. *The scope and method of political economy*. London: Macmillan.
Little, I.A.M. 1950. *A critique of welfare economics*. Oxford: Claredon Press.
Marshall, A. 1890. *Principles of economics*. London: Macmillan.
Mill, J.S. 1848a. *On the principles of political economy*. London: Longmans Green and Company.
Mill, J.S. 1848b. *The Principles of Political Economy (1947)*. Oxford: University Press.
———. 1859. *On Liberty*. London: Longman, Roberts and Green Co.
———. 1863. *Utilitarianism*. Indianapolis: Liberty Fund, 2013.
Myrdal, G. 1929/1932/1953. *The Political Element in Development of Economic Theory*, London: Routledge and Kegan Paul.
Putnam, H. 2002. *The Collapse of the Fact/Value Dichotomy and Other Essays*. Cambridge, Mass: Harvard University Press.
Ricardo, D. 1817. *On the principles of political economy and taxation*. London: John Murray.
Robbins, L. 1932. *An essay on the nature and significance of economic science*. London: Macmillan.
———. 1938. Live and dead issues in the methodology of economics. *Economica*, August 5 342–352,
———. 1981. Economics and political economy. *American Economic Review*. May: 1–10.
Rubinstein, A. 2012. *Economic Fables*. Cambridge: Open Book Publishers.
Sen, A. 1970. *Collective choice and social welfare*. Cambridge: Harvard U. Press.
Senior, N.W. 1951/1836. *An outline of the science of political economy*. New York: Augustus M. Kelly.
Sidgwick, H. 1883. *Principles of political economy*. London: Macmillan.
Smith, A. 1759. *Theory of Moral Sentiments*. New york: A.M Kelley.
Su, H., and D. Colander. 2013. A failure to communicate: The fact/value divide and the Putnam/Dasgupta debate. *Erasmus Journal for Philosophy and Economics* 6 (2): 1–23.

Chapter 7
New Theoretical City or Dispersed Tribes? An Exploration Journey through Contemporary Heterodox Economics and Methodology

Carlos Hoevel

Abstract The paper offers the reader what could be synthesized as a "fresh look" on the authors, currents and debates of heterodox economics and methodology. The intention is to present a non-pretentious and, if possible, entertaining pedagogical text for the use of non-experts and for the possible reflection of the experts. Following the best tradition of literary, philosophical and theological travel characters – such as Homer's Odysseus, Dante's Virgil, or Calvino's Marco Polo – the paper develops the theme on the journey story of "Adam B", an imaginary epistemological explorer of contemporary economics' territories. This traveler, like a cultural anthropologist or an adventurer, discovers new economic regions with no more elements in the backpack than his desire to know, some general pressing questions and a few specific research goals. The paper presents a series of descriptions, impressions and questions about the activities of the heterodox inhabitants - economists and methodologists - of economic lands.

In this paper I offer the reader what could be synthesized as a "fresh look" on the authors, currents and debates of heterodox economics and methodology. My intention is to present a non-pretentious and, if possible, entertaining pedagogical text for the use of non-experts and for the possible reflection of the experts. This text also aims to reflect as much as possible the different positions on the current state of economic science that are included in the rest of this book. Although I do not mention all the authors that participate in this volume, I hope that in the view that I will show, they can feel represented in some way. In that sense I ask for consideration for the errors in my interpretations and for the nuances and specific characteristics of the various positions that I will surely omit. My aim is not to analyze each position in depth, but to see them all in their mutual interrelation within a more general framework.

C. Hoevel (✉)
Pontifical Catholic University of Argentina, Buenos Aires, Argentina
e-mail: carlos_hoevel@uca.edu.ar

© Springer International Publishing AG, part of Springer Nature 2018

P. Róna, L. Zsolnai (eds.), *Economic Objects and the Objects of Economics*, Virtues and Economics 3, https://doi.org/10.1007/978-3-319-94529-3_7

Following with no talent the best tradition of literary, philosophical and theological travel characters – such as Homer's Odysseus, Dante's Virgil, or Calvino's Marco Polo - I will develop the theme of this article based on the journey story of "Adam B", an imaginary epistemological explorer of contemporary economics' territories. This traveler, like a cultural anthropologist or an adventurer, launches to discover new economic regions with no more elements in the backpack than his desire to know, some general pressing questions and a few specific research goals. I will rely on his trip diary from where I will take a series of descriptions, impressions and questions about the activities of the heterodox inhabitants - economists and methodologists – of economic lands.

Who is Adam B and why did I choose him precisely as a guide for this exploration? It is said that he received that name from his fervent desire to know, his ingenuity and his capacity for wonder similar to that of the first father of humanity. Besides, his name probably reflects his founding ambitions, such as those of the father of political economy. In a word, he seems to be a "B" version of both. This condition of being a curious, intrepid and at the same time non-expert explorer has its pros and cons. Among the obvious drawbacks I can mention the fact that he does not fully know the terrain, does not differentiate well its different inhabitants and does not have a clear position in relation to their struggles and disputes. Among the advantages, the main one is perhaps his naivety: Adam B approaches the new lands he crosses with more wonder, distance and perhaps more capacity for understanding than many of their oldest inhabitants.

As he writes in the introduction to his diary, the main motive that led Adam B to make this journey was the desire to know the roots of the problems of the world about which he suspects that the inhabitants of the economic territories have something to do. Besides this main purpose, I found among Adam B's notes the following preliminary questions: what happens with the great mainstream city of economics that, despite having been for so many years the model and guidance of world affairs, is today in the midst of the crossfire of so many critics both inside and outside the said city? Is this city the culprit, as many argue, of many of the problems that the world is facing today? And in the event that the latter is proven: would it be possible or necessary to reform partially or even totally this city? Besides, his diary refers also to three more specific research questions: (i) What are the different heterodox peoples that inhabit both inside and outside the territory of contemporary mainstream city? (ii) Which are their problems / challenges / roles / expectations? and (iii) Are they being successful in building a new theoretical city, or are they turning economics lands into a series of dispersed tribes?

What follows - that is, the rest of this essay devoted to Adam B's journey - is divided into four parts. In the first one, I will present Adam B's findings in the urban territory of the great mainstream neoclassical city and his observations on its heterodox settlers. In the second, I will refer to the inhabitants that he encountered when visiting the territories outside the neoclassical city grouped in the most important historical tribes of heterodox economics and some new ones about which our traveler gives also his personal point of view. In the third section, I will discuss the description and opinions of Adam B on the strange inhabitants of a third, very new, but also very prosperous territory: that of the experimental and empirical suburbs. Finally, and before my own conclusion, I will present some of the questions, positive possibilities and crossroads that Adam B raises especially on relation to the capacity of the inhabitants of the three economic territories to build a new theoretical economic city.

7.1 The Urban Territory of the Mainstream: Methodological Critics, Benevolent Interpreters, Theoretical-Experimental Critics, Performative detectives and Saboteurs

The first territory recorded by Adam B in the first stage of his journey is the very neat, organized and well-built urban territory of the mainstream, at the centre of which stands the great citadel of neoclassical theory. In addition to the many orthodox inhabitants who still live comfortably in their elegant buildings, Adam B informs us that he has found a large presence of more or less heterodox inhabitants. A first group of these settlers, some very old and others already living in their venerable tombs, is that of "methodological critics". These taught for many years that the main defect of the neoclassical city was not so much its content but its weak methodological construction. Thus, they devoted their life to submit many of the mainstream city buildings to a complex procedure – consisting in placing scaffolds, breaking walls and opening new spaces– that they called with the strange name of falsifiability. Their purpose was not, however, to weaken the buildings but to strengthen them in order to prevent their destruction by the external inhabitants of the city dominated, according to them, by crazy projects and utopias.[1]

Another group of more recent heterodox inhabitants found by Adam B are the so called "benevolent interpreters." These in principle support the realist claims of members of other heterodox tribes on the need for the city of economics not to remain isolated from the rest of the surrounding reality.[2] However, they do not share

[1] Douglas Wade Hands mentions Mark Blaug as the chief representative of this "tribe" of methodological critics: "Armed with the recent work in philosophy of science, as well as the history of economic methodology, Blaug makes a methodological evaluation of the conventional wisdom of economic science (...) He states in summary that his 'own contention...is that the central weakness of modern economics is, indeed, the reluctance to produce the theories that yield unambiguous refutable implications, followed by a general unwillingness to confront those implications with the facts' (p. 254) (Wade Hands 1993: 31–32).

[2] For example, Uslaki Mäki, one of the most brilliant representatives of this tribe, defines his realistic position as follows: "The world has an objective structure that is not created by scientists as they create their theories and models about that structure. Those theories and models are true or false in virtue of the ways of that objective structure – not in virtue of whether evidence supports them or whether we are otherwise persuaded to believe in them, for example. Finally, good science pursues theories that are true, while being prepared for the possibility of error" (Mäki 2007: 74).

the view of those who believe that abstract neoclassical buildings are a necessary indication of this unhealthy isolation. On the contrary, these benevolent interpreters, without moving a single brick, have devoted themselves to demonstrate almost a contrary thesis. According to them, the stylized and seemingly unadapted mainstream constructions - with their one-dimensional mirrors that deform or simplify everything that is around them - have a much greater capacity to reflect the complex reality than what many so-called critical realists believe.[3]

A third group of people, whom Adam B calls "theoretical-experimental critics", seem to be engaged in a series of new and strange projects. In fact, they seek to reform strongly both the foundations and the interior of several important neoclassical city buildings, but without modifying for the moment their external mathematical appearance. They are not, certainly, mere interpreters, but genuine reformers even if they wish to change the content without completely modifying the form.[4] Besides, the latter can also be seen excavating the earth around the great buildings, going back and forth along with some experimental micro-entrepreneurs arrived from other parts of the city and bringing new materials and tools that they introduce in the old buildings.[5]

In the same mainstream city Adam B also claims to have found other heterodox inhabitants: the performative detectives[6] (Ruggins 2013). Like the benevolent interpreters, they are more concerned with analyzing existing buildings than with changing them. However, they have a totally different intention to theirs. Being mainly detectives, they are true masters of suspicion who feel responsible for bringing to light the hidden mechanisms of neoclassical buildings. According to them, the orthodox mirrored skyscrapers, far from being just innocent elegant buildings designed to reflect the surrounding landscape, hide secret performative devices. By means of the latter, the neoclassical city has become an imperialist machine. Characterized by its constant expansion, this city does not receive passively the influence of the outside world but, on the contrary, it creates, as it were, its own environment.[7] For the rest, performative detectives do not seem to be definitely in

[3] As Uslaki Maki argues, "Economists build (or should build) theoretical models as representatives of real systems, as surrogate systems, the properties of which are *directly* examined in order to *indirectly* acquire information about real systems. Models are built and studied because there is no epistemically reliable "direct" access available to some deep facts of economic reality. All access is necessarily indirect and mediated by simple images of complex things. (...) What appears to be a model treated as just a freely floating fictional substitute system may be defended as a surrogate system by arguing that its connections to real systems are just very indirect or that those connections will be created in a long enough run" (Mäki 2007: 77).

[4] A notable representative of this group is, in my opinion, Robert Sugden (2000).

[5] These kind of procedures are described and supported by Donato Rodríguez & Zamora Bonilla (2009).

[6] I would mention Donald MacKenzie (2007) and Michel Callon (2007) as conspicuous members of this tribe.

[7] "The shaping of economies by economics –argues Donald MacKenzie-can be viewed as a triumph for the truths discovered by the discipline, or it can be condemned as the damaging of an abstract and unrealistic worldview; such matters remain fiercely controversial... The issue that needs to be tackled in relation to economies and economics is not just about 'knowing' the world,

favor or against the mainstream city. They are mainly devoted to the task of demonstrating that this metropolis is not intended to be a gazer to contemplate reality, but a living performative machine designed to produce it.[8]

Finally, at the last minute of his stay in the city, after nightfall, Adam B reports that he has seen, coming out of the sewers of the streets, some of whom he calls "external saboteurs". These intrepid incursionists do not dwell in the mainstream city but come from abroad. They visit the metropolis from time to time to remind its population that there is a world outside that they will never be able to ignore completely. Their task is bold and risky: they arrive in the evening, they enter the pipes and underground ducts, they look for the deepest and basic foundations of the great neoclassical buildings and they place an important but limited load of philosophical dynamite. Their aim is not to destroy the city nor to eliminate its inhabitants, but to drill and weaken its deepest foundations and convictions.[9] It was precisely when he was running after one of these intrepid saboteurs, that Adam B ended up outside the mainstream city limits and went out to the open, wild and endless plain of outer heterodoxy.

7.2 In the Wild Forests of Heterodoxy: Ancient Tribes, New Nomadic Groups and a pro-Independence Project

The second territory visited by Adam B, located outside the mainstream city limits, is inhabited by the largest and most varied old and new heterodox tribes. These dwell lost in an unending horizon of vast theoretical forests and endless interdisciplinary plains. Among the old heterodox, our traveler found representatives of the great theoretical tribes: Marxists, institutionalists, Keynesians, Austrians, Sraffians, historicists. Some of them once formed part of the mainstream city and even competed for its government. Now they live far away from the city limits and are considered extra-urban communities. Being their current territory a completely open landscape with no clear limits that mark its beginning or its end, these different

---

accurately or not. It is also about producing it. It is not (only) about economics being 'right' or 'wrong' but (also, and perhaps more important) about it being 'able' or 'unable' to transform the world" (MacKenzie, D. et al., 2007: 2). An excellent example of this, very well studied by MacKenzie, was the case of the Black-Scholes model: "By exploiting any discrepancies using the Black-Scholes model –argues Sara M. Ruggins- traders made the model *real* instead of a hypothetical representation of reality. Thus, Black-Scholes appears to be performative because it confirms the Barnesian thesis that 'the use of an aspect of economics alters economic processes so that they are more like their depiction by economics.'" (MacKenzie, D. et al., 2007, 77). Against these performative arguments, the reaction of the "realist" Mäki has been strong: "Instead of *economics performs the economy* as the sociologists say, it would make more sense to say, *the economy performs economics*" (Mäki 2013, 10).

[8] "Neoclassical price theorists care about the world in a different way. They do not want to alter their system to make it like the real world; they want to alter the real world to make it perform according to their system" (Mitchell 2010: 193)

[9] I would include in this group, for example John Davis (2006), certainly a peaceful man, but sharply critical of the philosophical assumptions of the mainstream.

scientific tribes can either remain isolated or linked almost at will, remaining completely free from all epistemological control of the authorities that govern the city.

In the same wild territories Adam B also discovered other new tribes bearing innovative, unusual and sometimes disconcerting names: feminists,[10] evolutionists,[11] Aristotelians,[12] personalists,[13] Buddhist,[14] civil economists,[15] etc. Many of their members are immigrants coming from the extra-economic territories of philosophy, anthropology, sociology and other sciences. Perhaps because these are relatively small groups recently settled in the territory, they tend to negotiate or make pacts individually or in small groups among them or with both heterodox and even orthodox people living outside or inside the city. In addition, Adam B reports the existence of the so-called pacifist travelers who, living most of the time in the territory of the outer heterodoxy, they move periodically to the mainstream city without feeling bound to adhere to any heterodox independence.[16]

Finally, Adam B found a new and powerful tribe completely at the antipodes of this lack of commitment to the independence of the heterodox territories. This new group formed by the so-called "critical realists" has shocked in the last years the different economic territories beyond and within the limits of the mainstream city (Lawson 2003). Their members are, above all, dedicated to achieving complete independence from the mainstream city. Its main banner is "ontology", which has only the name in common with the ontological flag raised by the aforementioned "benevolent interpreters". For the latter, the realistic approach involves considering the mainstream neoclassical city as a place built in an odd way but quite adapted to surrounding reality. Instead, for critical realists, neoclassical city is a deformed and abstract construction, completely removed from reality.[17] Thus, they consider that

---

[10] Cfr. Barker and Kuiper (2003).

[11] Cfr. Hodgson (1993).

[12] Some economists and epistemologists with clear Aristotelian influences are Amartya Sen (1987), Robert Skidelsky (2012), Arjo Klamer (2017), Peter Rona and Laszlo Zsolnai (2017), Ricardo Crespo (2013) and Deirdre McCloskey (1990).

[13] Such as for example the philosopher Luk Bouckaert (1999). I myself feel attached to this line of thought (Hoevel 2011).

[14] I am thinking in heterodox critics of neoclassical economics with strong Buddhist roots such as the business ethics expert Laszlo Zsolnai (2017).

[15] The main representatives of the civil economy school are the economists Stefano Zamagni and Luigino Bruni (2007).

[16] A typical pacifist traveler who explores with complete freedom all the territories of economics, including the mainstream is, in my opinion, the Italian economist Stefano Zamagni.

[17] Taking as an example their analyses of Austrian economics, Duncan Hodge shows the differences between the two conceptions of ontological realism of Mäki and Lawson: "According to Mäki, the Austrian emphasis on theory shows a realist orientation in the sense of explanatory redescription as, 'It is only by means of the conceptual resources of a theory -not being reducible to the observational language of empirical facts and generalizations –that empirical facts can be redescribed in a way which reveals what those facts really are.' (ibid. 321). (Hodge 2007: 29) "Lawson, however, finds much of Menger's methodological pronouncements to be deficient from a transcendental realist perspective... For Lawson, the fact that Austrian economics may be consistent with aspects of common sense, methodological, referential and veristic realism does not

its final destiny must be destruction or simply abandonment. Their project seems to be, according to Adam B's report, to found a New Ontological Independent Economic City: a daring initiative in these times when the great and old heterodox tribes look isolated and stripped of their past grandeur. This new city, however, would not have the same characteristics of the mainstream city. Unlike the closed, uniform, static, atomized and isolated character of the latter, critical realists imagine their new ontological city as an open, plural, dynamic, relational territory, in constant construction and fully integrated with both the rest of the heterodox territories as with other scientific territories. In addition, critical realists believe that this new ontological city will allow the creation of a new pluralistic space in common for peaceful coexistence and at the same time united progress of all heterodox tribes.[18]

7.3 In the New Suburban Neighbors of Experimental and Empirical Research: Experimental Practitioners, Inductive Developers and Neoclassical Explorers

Adam B was almost resigned to leaving the territories of contemporary economics with a feeling of unease due to the climate of fragmentation and division, when he gazed afar off some strange buildings that stood out in the suburbs and in some inner quarters of the distant city. In stark contrast to the grey, glazed and uniform buildings of the central neoclassical citadel, these triangular and colorful buildings attracted a lot of the attention and cheered the somewhat upset spirit of our restless explorer. Thus, instead of leaving, he decided to return to the city to discover the inhabitants of a new and unknown territory that he had not come to visit in his first stay.

This third territory is inhabited by a somewhat chaotic combination of new neighbors, swarming here and there, formed by people such as neuro-economists,

---

ensure that the way it represents the world is adequately realistic. From a critical realist position such as Lawson's, the meat of the issue is precisely how Austrian theorizing represents the particular target subsystems of social reality and whether it provides an adequate resemblance thereof. Lawson might not object to the idea that the Austrian representation of the norm captures an essential entrepreneurial aspect of its nature and may thus be regarded as realistic in this respect. However, it is evident from Lawson's account of Menger's approach to theory that he finds idealizing representations depicting individuals as 'omniscient' and 'infallible' to be a gross distortion of social reality in which the fictions being entertained are incapable of providing a satisfactory explanation of how the world really operates" (Hodge 2007: 30).

[18] "I have suggested, then –argues Tony Lawson- that the various heterodox traditions can be identified as heterodox through a recognition of the fact that they advance claims or practices or orientations which are either concrete manifestations of, or presuppose for their legitimacy, a social ontology of the (seemingly coherent) sort set out above. *In short, the set of projects currently collected together and systematized as heterodox economics is, in the first instance, an orientation in ontology*." (Lawson 2006: 498)

behaviorist practitioners, complexity theorists or computational researchers.[19] Due to the way in which these inhabitants move freely and without prejudices both inside and outside the neoclassical city, our man was able to verify the current porosity of the once well-delimited boundaries between the mainstream city territories and the outer heterodox lands. In fact, even though the neoclassical citadel still looks strong, the presence of these new dwellers demonstrates that it is no longer a completely homogeneous space. These suburban inhabitants penetrate the mainstream city everywhere, almost to its very centre (Colander et al. 2004).

In fact, many of the findings of these settlers devoted to experimental research, strongly question some of the central foundations of the orthodox citadel and introduce questions and ideas that until recently only had place in the outer heterodox territories. However, neither the economists involved in these investigations nor the methodologies they use are neither ascribable to any of the great traditional heterodox theories nor are they considered heterodox by the mainstream members themselves. Despite this, their new practices and fields of research, neither strictly orthodox nor truly heterodox, have almost completely captured the attention both within and outside the mainstream city, leaving relatively in the shadow the great theoretical projects of conventional heterodox traditions.[20]

On the other hand, beyond all their internal differences, among all these new heterodox neighbors there seems to be a common methodological strategy: pluralism (Colander 2007, Garnett Jr 2006). The latter allows them to combine a partial

[19] "In my 2001 book *Refection Without Rules* –writes Donald Wade Hands- I argued that economic methodology was moving away from the 'shelf of scientific philosophy' and more in the direction of naturalism, context-specific inquiries, and research that draws on a wider range of intellectual resources than just the philosophy of natural science. That process was ongoing at the time and has surely continued, but what was not clear a decade ago is how changes in economics itself have also initiated changes in the way that economic methodology is done. The bottom line is that almost all of the real 'action' within contemporary economic methodology is in precisely the fields that Davis and others point to as elements of the new, more pluralistic, mainstream: neuroeconomics, experimental economics, behavioral economics, evolutionary economics, and research employing new tools such as complexity theory, computational economics, and agent-based modeling. Neoclassicism may not be dead, but it is no longer the focus of the cutting edge of methodological research – but then nor is heterodox economics. Neither Neoclassicism nor Heterodox economics are the main focus of recent methodological inquiry." (Wade Hands 2012: 29)

[20] As Wade Hands argues, "this is not to say of course that Institutionalist economics, or Marxist economics, or other heterodox schools have completely disappeared, but simply that while there are many economists critical of mainstream neoclassical practice, those who are, seem to be focused on particular problems, applications, and tools, rather than self-identifying with any general heterodox school of thought. (Wade Hands 2012, in Lazzarini & Weisman: 21–22) "Notice that the vast majority of these books focus on experimental economics, but they all examine the economic research in one or more of the new more pluralistic microeconomic fields. Also notice that they all focus on either one particular field, or a small set of fields, within areas of economics that are neither heterodox nor strictly neoclassical. From a methodological perspective they are relatively close- focused studies: only certain aspects, authors, and applications within a field or small set of fields. These are also books with a normative philosophical focus – they are not at least primarily historical or sociological; they are philosophical – but again, it is a local or micro-philosophical focus, not the universal 'one rule fits all science' approach of earlier methodological work like that of Blaug and Hutchison." (Wade Hands 2012: 29)

acceptance of some valuable elements of the mainstream city (such as a certain Aristotelian substantialism that assumes the existence of common sense concepts like individual human beings or enterprises, which are not absorbed by an undifferentiated holism) with the new experimental findings that put into question other assumptions (such as utility maximization or a too strict methodological individualism). At the same time, within this new suburban heterodoxy, many share the strategy of not completely getting out from mainstream mathematical buildings (Sugden 2000). Rather, they seek to expand them with new mathematical facilities with the aim of endogenizing the findings that the new suburban researchers are bringing to light and that orthodox economists leave aside. What is common to these last positions is the conviction that the new experimental findings implicitly contain the theoretical building reforms that the great heterodox traditions had never been able to introduce. For this reason, many methodologists that travel between the three different territories consider that it is necessary to capitalize heterodox experimental findings by promoting, through a mathematical methodological adaptation, their incorporation into the very centre of the mainstream city (Colander 2007).

Other methodologists, especially those coming from the open heterodox plains, believe that in order to capitalize positively the achievements of the new experimental suburbs, it is necessary to avoid their absorption within the mainstream city. In order to do so, they strongly criticize the partial or complete integration of the suburb inhabitants within the mathematical theoretical buildings. Through various sophisticated strategies - such as translating whole methodological buildings from other sciences such as psychology or sociology – they try to connect the new experimental suburbs with the great heterodox theoretical tribes and not with the city (Lawson 2012). Finally, some economists still living in the city but critical of many of its most orthodox inhabitants, are completely skeptical about the true heterodox intentions of the new neighbors of these suburbs. Therefore, they prefer to dedicate their efforts to introduce heterodox issues (like happiness, intrinsic motivation, etc.) in conventional empirical research buildings and leave to their fate what they consider falsely progressive suburban people (Frey 2009).

7.4 After the Journey: Questions, Positive Possibilities and Crossroads

After reporting everything I read in Adam B's diary about what he saw and found in the aforementioned three economic territories, I would like to synthesize briefly some of his questions and opinions about their different inhabitants, the positive possibilities he finds in their projects and the final crossroads he considers that they are obliged to overcome:

7.4.1 Questions and Opinions

About the Inhabitants of the mainstream city:

(a) mainstream critical methodologists: Adam B notes that while these still have an important role, their time of central leadership among heterodox people seems to have passed;
(b) benevolent interpreters: about them our explorer asks: is it really possible to reform the mainstream city based solely on a benevolent-realistic interpretation of neoclassical constructions without touching almost nothing of their theoretical buildings that are so clearly non- realistic?
(c) theoretical-experimental critics: Adam B wonders if it is possible to change the interior of theoretical buildings – as the theoretical and experimental critics seem to think – without changing the form – that is, the mathematical structure;
(d) performative detectives: he asks if they are really true heterodox people or just part of orthodox imperialist troops;
(e) saboteurs: he thinks that these people are really outsiders although ultimately not very effective.

In a word, Adam B's main impression about the urban heterodox is that, despite all their criticism and reformist attempts, they do not seem to have any intention of leaving the mainstream city at all. Therefore, even if some of them may be willing to tear down some important walls of the neoclassical citadel, their project is by no means to join the outer heterodox tribes, but at all events, to stay inside a very superficially reformed mainstream city.

About the Inhabitants of the outer heterodox lands:

(a) traditional heterodox tribes: Adam B's main impression is that they are currently very aware – except for a few cases – of the fact of having become definitive exiles of the mainstream city and that they bear the expectation of surviving as such in the future;
(b) new nomadic groups and supporters of the pro-independence project: according to our traveler, most of these new heterodox seem convinced of the need to build a new type of theoretical city not separate or mathematical but interdisciplinary, realistic, relational and practical. However, the activities and goals of these new heterodox also raised in Adam B many doubts and questions. For example, with regard to the much-vaunted pluralism of its present inhabitants, Adam B finds this a good starting point but wonders if it can also be a good point of arrival. Other doubts have to do with some concepts raised by critical realists. For example, that of relationality. What does it mean? Of what kind of relationality are we talking about?[21] And as for the word "ontology": how difficult it is for

[21] I would add: Are we talking about relationality understood in a Hegelian or in a Buberian sense? In fact, I think that two very different economic cities would emerge from these two kinds of relationality.

our explorer to interpret it! Is it a method, a scientific theory or a philosophical framework? And finally, a question in which Adam B constantly insists in his travel diary: will the project of an ontological city –based on a visceral rejection of the whole theoretical structure of the mainstream city buildings – not end up pulverizing the very idea of any theoretical city for economic science, dispersing it in a series of indefinite and uncertain pluralistic and relational practices? (Hargreaves Heap 2004, Hausman 2000, Hodgson 2004).

About the Inhabitants of the new experimental and empirical suburbs:

About these people Adam B also gives rise to some difficult questions such as the following:

(a) will these new neighbors, to the extent that they continue to carry out their research within the space of mathematical theoretical buildings of the neoclassical city, become finally re-absorbed by the orthodox citadel of the mainstream territory?
(b) given that the activities carried out by these people are largely based on a purely naturalistic and fragmentary type of research, can they finally become isolated inhabitants devoted to simple empirical practices lacking a true theoretical structure and thus incapable to build a new theoretically founded economic city?

7.4.2 Positive Possibilities

After raising these very difficult questions I found the following paragraphs in which Adam B opens the way to positive possibilities. Let us see: "Although an exploration such as the one we have attempted generates all these questions, it is also true that from our observation of the three great territories of economic science – mainstream, heterodox and experimental – and of their inhabitants, it is also possible to glimpse some paths of collaboration between them. I dare to outline two possible ways in which I believe that things could proceed positively toward that goal:

(a) The first, in my view, would be a further growth of the intermediate territory of experimental research suburbs, which could gradually become a strong link between the mainstream city and the rest of heterodox lands. This would allow an even greater and fluid dialogue between the orthodox, the "urban" heterodox internal to the mainstream and the external heterodox of the "outer lands."
(b) A second, much more ambitious scenario could arise in the case of a gradual integration of current theoretical and experimental research suburbs into an ontological, relational, and practical New City that retains some positive elements of the neoclassical City combined with other heterodox tribes and other tribes outside economics."

7.4.3 Final Crossroads

"However –continues Adam B in a in a series of rather fragmentary long paragraphs that I also found- both scenarios would imply, in my opinion, the overcoming of at least three crossroads faced today by the heterodox inhabitants of economics: empirical-deductive, instrumental,[22] methodological realist,[23] ontological. The pluralist crossroads: certainly the new experimental opening of the neoclassical city is based on a flourishing methodological pluralism arising from a fruitful interdisciplinary communication with other scientific territories. Behind this methodological pluralism, a theoretical pluralism seems to emerge as well (Mearman 2010). My view is that both pluralisms are positive. This opinion is due, among other things, to the fact that an economic theoretical city should have its purpose in giving shelter to concrete, living and acting human beings that can never be reduced to the parameters of a completely regular and uniform city planning.[24] The complete

[22] "Constructive empiricism gives science the task of providing true descriptions of that which is observable – the ultimate aim of scientific theorizing is to construct models which are empirically adequate in this sense. As regards the hidden entities, essences and causal mechanisms postulated by such theories, the constructive empiricist does not deny their possible existence but suspends any judgment about them. Explanation is relegated to the realm of pragmatic considerations about the many different applied contexts or settings in which the question Why? may be asked." (Hodge 2007, 4)

[23] "For scientific realists, knowledge of some aspect of the world is unsatisfactory if it is confined to description and prediction. In addition to the questions: what is this like? and what will happen if?, such realists want to know why things are the way they are or happen the way they do. Scientific realism usually requires true explanations to be in terms of theoretical objects that are not directly observable in the sense that we have no immediate sense impression thereof. Such realists want to know, and believe it is possible to know, what is going on beneath the surface appearance of things. Such realists hold that theoretical objects such as viruses, genes, electrons and gravity exist (or refer to existing things), even though they are not directly observable – and that the inherent nature of such entities explains the various manifestations of things like illness, intelligence, electricity and falling apples. A radical scientific realist holds that only the objects of a scientific theory exist (or refer) whereas a radical common sense realist holds that only our ordinary unscientific notions of things exist (or refer). Most realists fall somewhere in between these two extremes." (Hodge 2007, 3–4)

[24] However, this does not mean that a theory of economic action is not possible at all. In fact, the goal of economics – as well as that of ethics – is precisely to achieve a theory of praxis (Crespo 2013). But this is only possible if one understands the theoretical complexity of the interaction between free action and conditioning/ regular structures. The basis of the need for this theoretical and methodological pluralism lies in the non-reducibility of human action to a strict causality. As pointed out by Scazzieri & Zamagni, for example John Hicks always defended the fundamental character of theory: "Hicks takes theories to be the product of a particular 'concentration of attention' (Hicks 1976a: 209). Theories are focusing devices that may be effective in bringing to view certain causal patterns, while leaving other (possible) causal patterns aside. This makes theories essential to economic analysis (as some concentration of attention is a necessary condition for the identification of a causal relationship)" (Scazzieri and Zamagni 2009: 1). However, they add, "in Hicks's conceptual framework, human choice is free and historical inevitability is rejected. This means that choice is seen as the ultimate determinant of actions, even if the actual outcome of any given choice reflects a 'structure' of possible events that is, to a large extent, independent of human deliberation (see Scazzieri 1993a, b)." (Scazzieri and Zamagni 2009: 3)

methodological uniformity of the theoretical buildings -as assumed by the current neoclassical orthodox inhabitants of the mainstream city– or its replacement by a new theoretical unity (sociological, psychological behaviourist, etc.) which would reproduce the same uniformity in new terms– is therefore, in my opinion, an inadequate solution (Gintis 2007).[25] Nevertheless, absolute pluralism does not seem to me a right position to adopt especially from a philosophical point of view. In this sense, I share the extended belief that integration between the various theoretical and experimental territories in economics could be expected, only through a kind of ontological, relational and practical common theoretical city. However, this would require, in addition to the theoretical and methodological pluralism within the territories of economic science, strong interdisciplinary relations with other scientific territories, especially the philosophical ones. Therefore, the theoretical unity of a new economic city would not be reached at the level of scientific theoretical buildings but at the philosophical level of wider ontological urban design. The latter could coexist perfectly with a strong theoretical and methodological pluralism at the level of scientific territories. This would certainly not imply a vertical subordination of the republic of economics to the kingdom of philosophy, but a type of philosophical republican collaboration that accompanies particular investigations.

(3) The dilemma between independence and dialogue: this is perhaps -as John Davis argues – [26](26) the most urgent and crucial crossroads for heterodox inhabitants -economists and methodologists- of any kind. Which of these two options will lead to good destiny not only to the latter but especially to the economic republic in general? The answer to this question –concludes Adam B- would give us the key to the final result of the journey of exploration that we have reported in this diary."

[25] "John Hicks was primarily a theoretical economist, but he never turned his interest in abstract concepts into one-sided attachment to any particular scheme of theory. He was, as Robin Matthews has noted, 'a conceptualiser' (Matthews 2004: 32; see also Matthews 1989). Indeed, he was 'more a toolmaker than a tool-user' (Matthews 2004: 32). He never allowed any particular point of view to conceal the variety of possible theoretical frameworks, however. In spite of having 'in his own mind a consistent system of thought' (ibid.), he was ready to accept the provisional and contingent character of specific economic theories. At the same time, he was convinced that theoretical schemes are essential to the understanding of economic reality. Hicks acknowledged the need for theoretical pluralism. He was not an eclectic economist, however. He adopted a pragmatic view of theorizing (Hicks 1985b, 1988). This led him to think that theories are context-dependent and that the switch from one situation to another may sometimes require the introduction of a different theoretical framework (Scazzieri and Zamagni 2009: 4). "Hicks, however, always preferred 'to combine elements from different theoretical systems rather than deduce his conclusions from a set of consistent hypotheses' (MacKenzie and Zamagni 1991: xxix) (Scazzieri and Zamagni 2009: 2).

[26] "If this is true, then traditional heterodox economists have two choices. They can maintain their outward-orientation, so that if change occurs in economics it will likely be on the terms determined by behavioral economists, experimentalists, and others in the new approaches. The risk here is that these movements may become more conservative as their success at influencing the core improves. Alternatively they can reverse their orientation, and turn to trying to shift what exists in the core, looking for allies in the 'new heterodoxy' along the way, so as to improve the chances of successful change for both." (Davis 2006: 28)

7.5 Conclusion: A New Theoretical City?

For several decades, heterodox economists and epistemologists have come to a more or less homogeneous consensus about the methodological and practical problems facing all the territories of economic science. One of these consensuses is that mainstream theory, despite having the pretension to reflect reality in a completely objective way, nevertheless seems more and more distant from reality. The ubiquitous theme in all the critiques of all the heterodox currents is, as everyone knows, the way mainstream theory reflects and even conditions the behavior of the real actors it describes. The theory of rational choice reduces this latter behavior to a scheme of efficient maximization between means and ends that critics consider – to my mind quite rightly – oversimplified. This conception concentrates unilaterally and punctually in analyzing the mechanics of calculations explicit or implicit in a supposedly efficient use of the means and inexplicably leaves aside the consideration of ends. It also leaves out multiple motivations other than maximizing the means. Finally, it ignores the complex cultural and social processes underlying the formation of the so-called preferences of the agents considering them as an exogenous data to the models. The root of this one-sidedness and inadequate simplification, argue most of these critics, lies in the dominant rationalism in the neoclassical core of mainstream theory, which induces economists to put aside the aspects of subjectivity, uncertainty, reflexivity, and complexity present in all human actions. These aspects are seen as purely subjective dimensions, belonging to the field of study of other sciences or capable of being considered only in practical action. However, since all these aspects are not completely measurable, they would not be part of economic theory (Blaug 2003, Hodgson 2012, Lawson 2012).

Now, although it seems quite clear from what place the heterodox authors – in their different and today much more varied types – want to leave, it does not seem so obvious the place where they want to arrive. What kind of knowledge would end up being this science or discipline or set of disciplines in charge of studying economic actions from this broader and more complex perspective that is proposed by many heterodox authors? Would it be a completely reformed economic theory or a partially changed intermediate new theory? In addition, would this new science constitute a separate discipline in itself? Or by leaving aside the ideal of a theoretical science – understood in the mathematical sense of neoclassical economic theory- would this new economic science also abandon any theoretical pretension in general? Would then it become a purely practical discipline? And if this happened, what kind of science would it be? Should economic science be a discipline without mathematical models, purely historical-hermeneutic, where there is an abundance of casuistic or managerial praxis, without any pretension to discover regular structures or laws and limited to individual practices or interventions, without any theoretical bases? Would not in this way economics become a completely fragmented activity, converted into a mere sum of contingent and relative actions that would lose all its weight as a scientific territory? And in the case that someone sought to avoid this drift by retaining the theoretical dimension in a key other than the mathematical / positivist, what kind of theory would it be? Would a sociological, cultural, political, or philosophical theory replace the current mathematical theory?

As we have tried to show in this article through the metaphor of the traveler in his journey across the territories of contemporary economics and methodology, the paths tried by the different heterodox currents to answer this difficult questions have been several. The finding of possible passages, although narrow and difficult to traverse, to leave this seemingly irresolvable questions and crossroads requires, in my opinion, a thorough elucidation of the concepts of theory and praxis and of the relations between the two. If by theory we mean only a kind of positivist and naturalistic knowledge for which the phenomenon of economic action can be unambiguously objectified and modeled, then an overcoming of the neoclassical trap would necessarily imply the complete abandonment of all theory. On the contrary, if the concept of theory is considered to be susceptible to analogical meanings, such as that of a non-mathematical theory, then one could think of a possible compatibility between the goal of moving away from neoclassical theory and of promoting a new economic science much more involved with praxis but without losing its theoretical dimension. In this sense, a next task would be to review among the outgoing paths proposed by heterodox economists - and also by orthodox ones– what elements could be detected that could be the basis for the constitution of such a discipline.

References

Barker, D.K., and E. Kuiper. 2003. *Toward a feminist philosophy of economics*. London/New York: Routledge.

Blaug, M. 2003. The formalist revolution of the 1950s. *Journal of the History of Economic Thought* 25 (2): 145–156.

Bouckaert, L. 1999. The project of a Personalistic economics. *Ethical Perspectives* 6 (1): 20–33.

Bruni, L., and S. Zamagni. 2007. *Civil economy: Efficiency, equity, public happiness*. Bern: Peter Lang.

Callon, M. 2007. What does it mean to say that economics is performative? In *Do economists make markets? On the performativity of economics*, ed. Muniesa MacKenzie and Lucia Siu, 310–357. Princeton: Princeton University Press.

Colander, D. 2007. Pluralism and heterodox economics: Suggestions for an "inside the mainstream" heterodoxy, Middlebury College Economics, Discussion Paper No. 07–24, pp. 1–11.

Colander, D., R.P. Holt, and J.B. Rosser Jr. 2004. The changing face of mainstream economics. *Review of Political Economy* 16 (4): 485–500.

Crespo, R.F. 2013. *Philosophy of the economy: An Aristotelian approach*. Dordrecht: Springer.

Davis, J.B. 2006. *"The nature of heterodox economics" Post-autistic Economics Review*. Vol. 40, 23–30.

Frey, B.S. 2009. *Economics as a science of human behaviour: Towards a new social science paradigm*. Dordrecht: Springer.

Garnett, R.F., Jr. 2006. Paradigms and pluralism in heterodox economics. *Review of Political Economy* 18 (4): 521–546.

Gintis, H. 2007. A framework for the unification of the behavioural sciences. *Behavioral and Brain Sciences* 30: 1–61.

Hargreaves Heap, S. 2004. Critical realism and the heterodox tradition in economics. In *Transforming Economics: Perspectives on the Critical Realist Project*, ed. P. Lewis. London, New York: Routledge.

Hausman, D. 2000. *"El realismo crítico y las teorías de sistemas abiertos" Argumentos de Razón Técnica*. Vol. 3, 61–92.

Hicks, J.R. 1976a. Revolutions in Economics. In *Method and Appraisal in Economics*, ed. S.J. Latsis, 207–218. Cambridge: Cambridge University Press.

———. 1985b. Sraffa and Ricardo – a Critical View. In *The Legacy of Ricardo*, ed. G.A. Caravale, 305–319. Oxford: Basil Blackwell.

———. 1988. Introductory remarks at the International Economic Association conference 'Value and Capital –Fifty Years Later,' mimeo, University of Bologna.

Hodge, D. 2007. Economics, realism and reality: A comparison of Mäki and Lawson, Working Paper Number 63, University of South Africa, 1–39.

Hodgson, G.M. 1993. *Economics and evolution: Bringing Life Back into Economics*. Cambridge: Cambridge University Press and University of Michigan Press.

———. 2004. Some claims made for critical realism in economics: Two case studies. *Journal of Economic Methodology* 11 (1): 53–73.

———. 2012. *"On the limits of rational choice theory" Economic Thought*. Vol. 1, 94–108.

Hoevel, C. 2013. Economics, business and the personalist philosophy. In *Spriritual Humanism and Economic Wisdom*, ed. H. Opdebeeck and L. Zsolnai, 23–34. Anwerp: Garant.

———. 2013. *The economy of recognition. Person, market and Society in Antonio Rosmini*. Dordrecht: Springer.

Klamer, A. 2017. *Doing the right thing: A value based economy*. London: Ubiquity Press.

Lawson, Tony. 2003. *Reorienting economics*. London/New York: Routledge.

Lawson, T. 2006. The nature of heterodox economics. *Cambridge Journal of Economics* 30: 483–505.

———. 2012. Mathematical modelling and ideology in the economics academy: Competing explanations of the failings of the modern discipline? *Economic Thought* 1: 3–22.

MacKenzie, L.W., and S. Zamagni. 1991. Introduction. In *Value and Capital Fifty Years Later*, ed. L.W. McKenzie and S. Zamagni, xviii–xxxix. London: Macmillan.

MacKenzie, D., et al. 2007. Introduction. In *Do Economists Make Markets? On the Performativity of Economics*. Princeton: Princeton University Press.

Mäki, U. 2007. Realistic realism about unrealistic models. In *Oxford Handbook of the Philosophy of Economics*, ed. Harold Kincaid and Don Ross. Oxford: Oxford University Press.

———. 2013. Performativity: Saving Austin from MacKenzie. In *Perspectives and Foundational Problems in Philosophy of Science*, The European Philosophy of Science Association Proceedings, ed. V. Karakostas and D. Dieks, vol. vol 2. Cham: Springer.

Matthews, R.C.O. 2004. Hicks, Sir John Richard (1904–1989). In Oxford Dictionary of National Biography: From the Earliest Times to the Year 2000, ed. H.C.G. Matthew and B. Harrison, vol. XXVII. Oxford: Oxford University Press (in association with the British Academy), 32–3 [online edition: www.oxforddub.com, accessed May 27, 2007].

McCloskey, D. 1990. *If You're so smart: The narrative of economic expertise*. Chicago: The University of Chicago Press.

Mearman, A. 2010. Pluralism, heterodoxy and the rhetoric of distinction. *Review of Radical Political Economics* 43 (4): 552–561.

Mitchell, T. 2010. The resources of economics: Making the 1973 oil crisis. *Journal of Cultural Economics* 3 (2): 189–204.

Rodríguez, de Donato. 2009. Xavier & Zamora Bonilla, Jesús (2009), credibility, idealisation, and model building: An inferential approach. *Erkenn* 70: 101–118.

Rona, P., and L. Zsolnai. 2017. *Economics as a moral science*. Dordrecht: Springer.

Ruggins, S. M. 2013. *On the performativity of economics* https://qspace.library.queensu.ca/bitstream/handle/1974/8360/Ruggins_Sarah_201309_MA.pdf;jsessionid=C8A362FA0F4172669870091F9062A207?sequence=1

Scazzieri, R., and S. Zamagni. 2009. Between theory and history: On the identity of Hicks's economics. In *Markets, money and capital: Hicksian Economics for the Twenty-First Century*, ed. R. Scazzieri, A. Sen, and S. Zamagni, 1–41. Cambridge: Cambridge University Press.

Scazzieri, R. 1993a. *A Theory of Production: Tasks, Processes and Technical Practices*. Oxford: Clarendon Press.

———. 1993b. Actions, Processes and Economic Theory. In *The Makers of Modern Economics*, ed. A. Heertje, vol. I, 84–114. London: Harvester Wheatsheaf.

Sen, A. 1987. *On ethics and economics*. Oxford/New York: Basil Blackwell.

Skidelsky, R., and E. Skidelsky. 2012. *How much is enough? Money and the good life*. New York: Other Press.

Sugden, R. 2000. Credible worlds: The status of theoretical models in economics. *Journal of Economic Methodology* 7: 1–31.

Wade Hands, D. 1993. *Testing, Rationality and Progress*. Lanham: Rowman & Littlefield.

Wade Hands, D. 2012. Orthodox and heterodox economics in recent economic methodology. In *Perspectives on Epistemology of Economics. Essays on Methodology of Economics*, ed. A. Lazzarini and D. Weisman, 17–42.

Part IV
Temporality, Reactivity and Crowding

Chapter 8
Rational Choice Theory and Backward-Looking Motives

Roger Teichmann

Abstract The paper argues that the philosophical underpinnings of rational choice theory are vitiated by consideration of the phenomenon of backward-looking motives, such as gratitude, fidelity, and many forms of honesty. Attempts to describe the actions and decisions of those acting from such motives in the terms of rational choice theory fail, and the model of human conduct which is implicit in the theory is both inadequate in itself and pernicious in its general influence. A picture may emerge of the human person as a repository of atomistic 'experiences', inhabiting a present from the vantage-point of which his relations to past and future 'selves' are contingent, and his responsibilities and commitments correspondingly fluid. And in economics, a supposedly 'neutral' account of rational choice in fact encourages and sustains a simple-minded utilitarian ethic.

There is a well-known model of rational agency according to which a rational person's reasons for doing X will be such that doing X is either his end, or is something he takes to be a means to his end (or one of his ends). According to this model, a person has as end or goal some achievable state of affairs, such as feeling pleasure, and this goal renders certain courses of action *prima facie* suitable, namely those courses which would bring about the state of affairs, or would raise the likelihood of its coming to be. The person may have false beliefs about which courses of action are thus suitable for him, but this merely indicates the need for a distinction, between what he has 'objective' reason to do, and what he has 'subjective' reason to do. His practical rationality will not as such be impugned if he does something which is only 'subjectively' rational, though his theoretical rationality may be impugned, if it is the explanation for why the relevant belief of his is false.

This sketch will be familiar, as describing a model of rational choice which is standard among economists and decision theorists, its key notion being that of 'expected value', or alternatively 'expected utility'. It is a sketch only, and more

This paper is based on an earlier paper by the author, 'The Importance of the Past' (*Philosophy* 88 2013, The Royal Institute of Philosophy). The publisher's permission to use the paper is greatly appreciated.

R. Teichmann (✉)
St. Hilda's College, University of Oxford, Oxford, UK
e-mail: roger.teichmann@st-hildas.ox.ac.uk

© Springer International Publishing AG, part of Springer Nature 2018
P. Róna, L. Zsolnai (eds.), *Economic Objects and the Objects of Economics*, Virtues and Economics 3, https://doi.org/10.1007/978-3-319-94529-3_8

detail could be added. But for present purposes, we need only look at a class of reasons for action which are inexplicable on this model of agency - namely, backward-looking reasons.

An example of a backward-looking motive is the motive of gratitude for a service done. As Elizabeth Anscombe (1957) brought out so effectively, reasons and motives are alike elicited by the question 'Why?', as in 'Why are you doing that?', and if I ask you why you are sending flowers to Emily, let us suppose you reply, 'Because she drove me to hospital when I broke my arm'. On the face of it, your reason mentions a past fact, a fact of recent history. But the philosopher who adheres to the 'expected utility' model will look for a way of describing your action that makes it appear to be aiming at bringing something about, and will thus reject 'Because she drove me to hospital' as a statement of the 'real' reason. What aim or goal could be alleged in this case? Two kinds of answer suggest themselves. The first kind makes use of the key word, here 'gratitude', and inserts it into the means-end format, thus yielding the putative reason for action: 'In order to show my gratitude'. The second kind of answer relies on the notion of pleasure, as a state that is produceable by actions; we then either have the egoistic reason, 'In order to derive pleasure from my own action', or the more altruistic 'In order to give pleasure to Emily'.

Both these accounts can be shown to be mistaken, and for similar reasons. Your sending flowers to Emily may in a sense be called a means to your end, that of expressing gratitude, insofar as it is a *way* of expressing gratitude. But we have to ask, 'What is gratitude?' And any adequate answer will allude to what sort of reason for an action makes it a grateful action - namely, such a reason as 'Because she drove me to hospital'. The force of 'In order to show gratitude', in other words, here rests on the prior force of a backward-looking reason. It is no good treating the sending of flowers to Emily as your goal or desired outcome, to which means-end deliberation is indeed relevant; for this 'goal' does not in fact give your ultimate reason for action. For you can informatively answer the question, 'Why are you sending flowers to Emily?', and it is your answer to this that gives your ultimate reason. The decision theorist may allege that the reason you thus give has the function of ascribing a 'value' to the outcome of Emily's receiving your flowers, in this way insisting that your goal is after all this outcome. But this 'value' is evidently not *intrinsic* to the state of affairs in question, since it would lack that value if Emily had never taken you to hospital. How might one explain to someone what this 'value' consists in? Answer: by employing or invoking the statement, 'Because she drove me to hospital'. The other person would understand your explanation of the value of sending flowers if and only if they were familiar with the force of such a statement as a reason for action.

It is worth stressing this point about the notion of 'value', since it is a feature of decision-theoretic discourse to present the notion as if it is fairly well-understood, and then to cash it out for us in terms of hypothetical *preferences*. Thus, it may be stipulated that the value you ascribe to Emily's receiving your flowers is equivalent to the smallest sum of money you would accept in lieu of her receiving the flowers. There are already problems arising from the possibility that someone doesn't regard money as desirable in the way that the theorist wishes him to; and it is likely to be absurd to claim that if a person would accept *no* sum in lieu of Emily's receiving flowers then he must ascribe infinite value to that 'outcome'. But even if you *would* accept some lowest amount of money in preference to sending flowers to Emily, this doesn't show that your ultimate reason for action, when you do in fact send her flowers, was a forward-looking reason. For the hypothetical preference in no way removes or translates away the reason, 'Because she drove me to hospital'. All that we can say is that another (forward-looking) reason could, hypothetically, outweigh the backward-looking one, as far as your deliberations went.

I said just now that the so-called value of sending flowers to Emily would not be intrinsic to the act; but some philosophers do speak of such actions as grateful acts and promise-keepings as having intrinsic value, as also such actions as ungrateful acts and promise-breakings, these latter having intrinsic negative value. Thus John Broome (1995: 4) writes, "If you break a promise, one consequence will be that you have broken a promise, and the wrongness of promise breaking can be taken as a bad feature of this consequence…In this way, the intrinsic value of acts can be absorbed into teleology." ('Teleology' is the form of utilitarianism Broome is defending.) If I keep a promise, then presumably it is under the description *keeping a promise* that my action is to be made out as having intrinsic value - for it probably won't have any intrinsic value, except accidentally, under some such description as *bringing Smith a book*. But one could argue analogously that under the description *providing for my descendants* the action of investing my money in a certain way has intrinsic value. This latter way of talking would usually be ruled out by distinguishing the value an act has in itself from the value it has on account of what happens subsequently (especially what happens as a result of the act); and similarly, one ought to distinguish the value an act has in itself from the value it has on account of what happened earlier. The crucial point to make concerning your sending flowers to Emily is just that, if we are to speak of the action's having a certain value, then that value depends on past facts, and as far as rational choice goes, it is these past facts that the agent must mention in giving his reasons for his action.

Once it is recognised that the ultimate reason for a grateful action is not given just by mentioning the action itself, the inapplicability of the expected utility model can be seen in the fact that there is no place for an assignment of *probability* in connection with the agent's ultimate reason. If sending flowers to Emily were my ultimate end, then to be sure I could assign some probability to my achieving that end, given my adoption of certain means, such as phoning Interflora. But I cannot sensibly ask, 'If I send flowers to Emily, what is the probability of my achieving my end, that of showing gratitude?' In sending the flowers I *thereby* show my gratitude. One seems forced to say that the probability of my showing gratitude is 1 – which indicates, not a special source of epistemic certainty, but the pointlessness of this way of talking. It is essential to the expected utility model that a rational agent both assign some value to a given outcome, and also assign some probability to it (given the adoption of certain means). So the expected utility model is inapplicable to actions, such as grateful ones, whose ultimate reasons are backward-looking, on account of the inapplicability of the notion of probability to those reasons.

What of the second sort of account I mentioned (five paragraphs back), the sort that relies on the notion of pleasure as a state produceable by one's actions? It can be asked what sort of pleasure it is that is allegedly being aimed at. Presumably any pleasure felt by the sender is meant to be pleasure in the fact of the recipient's pleasure. And Emily will feel pleasure on receiving flowers because flowers are lovely things; but why do you want to bring her this pleasure? If the reply is, 'Because she drove me to hospital', the backward-looking reason has resurfaced. If the reply is, 'To feel pleasure myself', then why choose Emily, as opposed to any other susceptible individual? Again, the obvious answer is, and must be if what we have here is gratitude, 'Because she drove me to hospital'.

A rather desperate move may be made here, that of bringing in induction, in the following sort of way: 'I have found in the past that the pleasure I get when I give pleasure to people who have helped me is greater than if I were giving pleasure to other kinds of people; so I now calculate that if I send flowers, etc. etc.' Now the idea that a type of intentional action should be *found* to produce the thing that constitutes its own rationale is clearly a bit peculiar. This is especially true of actions not of the sort that you were trained to do in childhood: for it could be alleged that actions of the sort that you were trained to do in childhood were initially motivated by something different, namely a desire for parental approval (or perhaps indeed these early actions were cases of mere imitation, not yet fully intentional). But consider the person who for the first time in her life performs an act of vengeance, this not being the sort of thing inculcated by Mummy and Daddy. What can the aim of such an action be, on this first occasion? There is, after all, no experiential basis for predicting any particular outcome, e.g. a 'pleasure state' - so it seems as if the action must be pointless, according to the 'expected utility' model. (We might now postulate some reason like, 'I have noticed that other people get pleasure from vengeful acts, so on the assumption that I am like them, I calculate….' - evidently ludicrous.) Apart from all this, there are in any case backward-looking motives which it is quite implausible to reduce to forms of pleasure-seeking, such as motives of fidelity or loyalty that result in self-sacrifice.

All in all, then, the case for the existence of genuinely backward-looking reasons looks rather strong. Such reasons appear to falsify the 'expected utility' model of rational agency, or indeed any model which insists that rational agents always look either to the future effects of their actions or to those actions' intrinsic value. But it may now be claimed that although we do give and accept facts about the past as reasons for our actions, this very fact shows our irrationality. Maybe certain tribes have given and accepted as a reason for not doing X that X is 'taboo'; but can't we criticise that practice as irrational? Mightn't it be the same with backward-looking reasons?

When we criticise the notion of 'taboo', used as I am imagining it being used, we do so on account of the fact that the term 'taboo' fails to link up to human good or ill, in the sort of way which might make its prohibitive force intelligible. 'Why should I avoid eating that?' 'You'll be sick if you do.' This answer is quite intelligible; and the further question, 'Why should I avoid being sick?', points to an imagined need for a reason where there is none. By contrast, if the answer to 'Why should I avoid eating that?' were 'Because eating it would be taboo', then no reason would yet have been given, unless the activities listed as taboo can be made out as harmful, or bad in some way.

The intelligibility, as a reason for grateful action, of 'Because she drove me to hospital' is connected to what is good or bad for human beings, just as is the intelligibility of 'Because you'll be sick'. And this is because the phenomenon of gratitude, or returning good for good, is normal and natural for human beings (like loving your offspring and resenting injury), but additionally because a prevalence of habits of gratitude is conducive to social harmony, and this fact makes a reason for action which is expressive of gratitude fitted *as* a reason, in the context of the social language-game of asking for and giving reasons. Our concepts of what a good person is, and of what is good for human beings, are not *purely* naturalistic: their shape is partly determined by the social and empirical nature of the human practice of reason-giving. Giving reasons for what you do is giving an account of yourself, especially to others.

I spoke just now of the habit of gratitude as being conducive to social harmony; and this may seem to suggest that we do, after all, have a forward-looking rationale or justification for the human practice of accepting backward-looking reasons. The statement 'This practice is a good idea because it will (continue to) achieve social harmony' surely invokes a forward-looking reason? But, firstly, we did not *decide* to institute the practice of having backward-looking reasons, nor are we continuing with this practice for any *reason* - any more than we speak a language because of the expected benefits of doing so; secondly, even if we did have forward-looking reasons for going in for the practice, this would not show that the moves within the practice, i.e. the giving and accepting of backward-looking reasons, were after all not what they appeared.

One could perhaps argue that facts about social harmony help to explain the survival and spread of this practice among human beings; but this would not reduce the force of backward-looking reasons to that of a forward-looking one, nor would it provide a 'justification' for the practice, any more than an empirical explanation for the success and spread of an empire with its customs *ipso facto* provides a justification, either of the empire or of its customs.

I said above that habits of gratitude, and similar habits, are normal and natural for human beings, a fact that helps to explain why backward-looking reasons count as compelling within the language-game of asking for and giving reasons for action. However, the human phenomenon under discussion is richer than the phrase 'normal and natural' would suggest, all on its own; for part of what is at issue is our way of perceiving the world, and ourselves and our actions. This 'way of perceiving' is above all non-atomistic, and our notions of what is *significant* are typically contextual and holistic.

Understanding the significance of something often involves seeing it against the background of a larger context, a bigger picture. One instance of this truth is Frege's Context Principle, as enunciated in the Introduction to *The Foundations of Arithmetic*, where Frege enjoins us 'never to ask for the meaning [*Bedeutung*] of a word in isolation, but only in the context of a proposition'. Another instance can be found within human institutions and rule-governed practices: a piece of paper only counts as a ten pound note against a background containing banks, activities of buying and selling, etc. etc. And many events and phenomena have the significance they do because of what led up to them, or came before them. For the 'background' to an event, insofar as it is something we can know about, talk about, agree or disagree about, will typically be in the past.

Take an event within a piece of music. A cadence, a climax, a broadening out, and numerous other kinds of musical event, are what they are because of what has led up to them. You only hear those two chords as constituting a modulation, and so as being a harbinger of peace, or a cloud moving across the sun, or whatever, on account of what has just preceded them. Now this is not to say that what happens *after* a musical event doesn't also help determine the character of the event. Thus the point of a gradual build-up is evidently in what it builds up to. But even here there is a certain priority of past over future: for the characterisation of the passage as a build-up relies on the perspective of the listener who finally hears the apotheosis, in the light of what *has already happened* - even though past musical experience may on occasion incline you to guess that what you're presently hearing is bound to be a build-up of some sort. (Rossini.) You can, after all, be cheated, and hence proven wrong, by a mischievous or incompetent composer; whereas a certain pair of chords cannot help constituting a modulation as they occur, given what has happened. What is written is written.

Analogous remarks can be made about the events of human life. Your sending flowers to Emily is a case in point: it is what it is because of what Emily has done for you in the past. Again, this is not to deny that a historical event will often assume a certain significance on account of what came after it - as with the assassination of Archduke Franz Ferdinand in 1914. But here too it is the perspective of those who recall the event which is crucial. For it is in effect they who give it its significance, which is thus a retrospective significance.

A possible partial diagnosis of why people are attracted to the expected utility model of agency suggests itself. Perhaps such people are imagining states of affairs as necessarily having some sort of intrinsic value, intrinsic significance. If they were right it would follow that the significance of a present action and a past state of affairs must be mutually independent, so that only the prospect of affecting 'overall significance' would seem to supply any reason for or against performing an action - which is to say, only the significance of the action itself, and/or the possible significance of future states of affairs, would seem to supply such a reason. But if I am right that there are species of significance that involve the broader context, and hence (in the case of events in time) that involve the past, then this notion of 'intrinsic significance' cannot be universally applicable.

It is even doubtful if it is ever applicable; for an investigation of those concepts that have, in the traditional accounts, been particularly associated with 'intrinsic value' shows, I think, that the traditional accounts have got things wrong. I am thinking especially of the concept of pleasure, which is of course the key concept for classical utilitarians, for whom the prospects of producing pleasurable states of affairs are what supply reasons for rational action. The idea that pleasure has intrinsic (positive) value goes with a picture of it as a sort of sensation or quasi-sensation, or in other words as a self-intimating episode of consciousness, with a certain duration, typically having a certain intensity, and with an intrinsic 'character' knowable authoritatively by the subject. But pleasure is no such thing; it is much more connected to the 'space of reasons' than the classical utilitarian realised. The question 'What's the pleasure in that?' is often a real one, and can receive answers embodying good, bad or unintelligible reasons. (For more on this topic, see Teichman 2011).

As regards the intelligibility of the reasons people give for their actions (past, present, or proposed), it seems clear that one source of this intelligibility will very often be how their actions fit into a bigger picture, involving other people, other actions, other times and places - all of which contribute to an action's 'significance', as I have been calling it. This term is not meant to denote some single intuitible property à la G.E. Moore. Rather, 'significance' is a portmanteau term covering a family of ordinary but crucial action-concepts, such as gratitude, fidelity, temerity, tact, aimlessness, solidarity....And an important sub-class of these action-concepts relate to backward-looking motives.

A theory of rational action and rational deliberation which ignores such motives, or presents a distorted account of them, is liable to have various ill effects, especially if it seeps into the culture at large, or becomes part of the ideology of those who govern us. For such a theory will present an impoverished picture of human beings and of human life. History comes to be regarded as at best a source of useful pointers as to how to get what we want; phenomena such as gratitude, fidelity, and justice become puzzling, and attachment to them can seem mere superstition. A picture may emerge of the human person as a repository of atomistic 'experiences', inhabiting a present from the vantage-point of which his relations to past and future 'selves' are contingent, and his responsibilities and commitments correspondingly fluid. (The late Derek Parfit received praise and honour for his compendious filling-in of this picture.) And in economics, a supposedly 'neutral' account of rational choice in fact encourages and sustains a simple-minded utilitarian ethic.

References

Anscombe, G.E.M. 1957. *Intention*. Oxford: Blackwell.
Broome, J. 1995. *Weighing Goods*. Oxford: Wiley-Blackwell.
Teichmann, R. 2011. *Nature, Reason And The Good Life*. Oxford: Oxford University Press.

Chapter 9
Time-Value in Economics

Kevin T. Jackson

Abstract Modern economics deploys the concept of time-value as if it were a descriptive, empirically-verifiable objective property of the sort studied by natural science in its investigation of the natural world. In particular, economics frames efficiency as a core feature of time-value understood in a purportedly descriptive sense. Economic time-value is stripped of its normative quality, and held up for scientific investigation and brute factual description of the sort associated with criterial and natural-kind concepts. As such, economics deploys the conceptual devices such as the "time-value of money" in its quest to objectively ascertain in a purportedly morally-neutral fashion behaviors that are favorable to capital efficiency maximization. Yet, as the paper shows, economics thus removes the relation of efficiency to time-value from the realm of normative choice – that is, from the realm of contested interpretive concepts.

Economics privileges a purportedly descriptive account of time-value and uses it as a mathematized conceptual device for depicting notions such as net-present value of money and efficiency, in ways that portray such concepts as factually determined and necessary, thereby imitating the laws of natural sciences. Yet such economic concepts that are constructed on the basis of seemingly descriptive, "value free" representations of time-value eclipse the often more significant normative character of time-value as comprehended from a wider perspective.[1] The prominence of the

[1] In this paper I am discussing the concept of time-value in a wider sense than its narrow use in finance and banking that refers to the market value of an option over and above its intrinsic value. See, e.g., Oxford Dictionary of Finance and Banking. Oxford University Press. The broader philosophical concept of time-value that I exposit in a larger research project is that time depends on value, and value depends on time. There can be no valuation (whether economic, aesthetic, or moral) without time. While one may have a sense of timelessness when contemplating a masterwork of sculpture, the contemplation takes place within a duration of time. Conversely, there is no time without valuation; what appears to be "descriptive" or "factual" valuation is ultimately "pre-

K. T. Jackson (✉)
Université Libre de Bruxelles, Bruxelles, Belgium

Fordham University, New York, USA
e-mail: kevin.jackson@ulb.ac.be

© Springer International Publishing AG, part of Springer Nature 2018

P. Róna, L. Zsolnai (eds.), *Economic Objects and the Objects of Economics*, Virtues and Economics 3, https://doi.org/10.1007/978-3-319-94529-3_9

reductive descriptive account of time-value in modern economics thereby masks the role that such an account plays in advancing a particular normative agenda. Economics brackets off a supposedly "value-neutral" construction of time-value that portrays as objectively necessary (as "law-like") a class of operations arranged at achieving "maximum efficiency" — that is, operations that incentivize "rational agents" to invest in short-term returns rather than pursuing alternatives such as longer-term growth, well-being, or other potentially more sustainable options.

The attention economics pays to time-value is largely restricted to the sphere of objective, measurable features of time which are, in turn, processed through mathematical formulae to empirical theories of human behavior based on psychological studies. The treatment of time-value by positive economics gets cashed out in the familiar axiom "time is money." This preoccupation of economics with only the factual and measurable side of time-value – what I shall call time-value-as-plain-fact – carries over into mainstream understandings of business, resulting in a variety of anti-social tendencies resulting from mindless short-termism, and a host of perverse injustices undertaken in the name of an assumed deterministic efficiency.[2] Time has become an increasingly important element in fueling competition, not merely in terms of the provision of goods and services more rapidly but also in unlocking new markets. Financial markets in particular increasingly assign greater weight to short-term performance[3] – not merely quarter-to-quarter or even month-to-month but millisecond-to-millisecond (Davis et al. 2012). From a theoretical vantage point, the upshot is a perpetuation of the pretense of economics as a value-neutral, descriptive enterprise.

9.1 Time-Value as Fact

Economics takes its understanding of time-value-as-fact from an underlying concept of time provided by physical science: time moves in one direction – the past lies behind, fixed and immutable.[4] The future lies ahead and is not necessarily fixed. According to the Second Law of Thermodynamics, natural processes have a preferred direction of progress. Thus, heat flows from area of higher temperature to

scriptive" and "normative." Our virtually unainimous adoption of clocks and calendars is ultimately a convention that expresses a preference for using discrete equal units of fixed duration for counting time. Strictly speaking, there is no "law" of nature that dictates this choice.

[2] "In reality the market lends more importance than is justified to the short-term future, ignoring the long-term damage this can inflict, particularly on relationships with workers, suppliers and customers" (Hutton and Jones, 2005 at 93).

[3] "Investment and careful building-up of market share are sacrificed to boost profitability in the immediate future. It is the share price now that matters, not the higher return that could be achieved in five years' time. In fact, the very structure of the financial economy prioritises short-term gain" (Hutton and Jones, 2005 at 92).

[4] Cf. Aristotle, "Even God cannot change the past." *Nicomachean Ethics*, Book VI, Sect. "Normative Dimension", 1139b.

area of lower temperature. It is never the reverse, unless some kind of external work gets performed on the system in question. Thermodynamic processes reduce the state of order of initial systems, so entropy is an expression of disorder or randomness.

Ultimately, entropy determines the direction of the "arrow of time." The gas combustion that makes a car run creates mechanical energy and heat. But no known process in the universe will let exhaust gases re-combine with the heat energy to reconstitute the original gasoline. The heat energy of the burning gasoline reaches a higher, irreversible state of randomization. The entropy of the system, and the "material" universe, is irreversibly increased, as required by the Second Law.

Shifting focus from physics to economics, in an important way, economic value is analogous to energy: it becomes randomized as with the Second Law. For instance, spending money is easier than earning, saving, and investing it. Yet once created, economic value requires infusions of energy from sources outside a system – e.g., the energy of competent strategy; the energy behind contributions of capital or labor.

Indeed, without temporally ordered management able to organize and structure transactions over time, a business will fail, a victim of entropy. To give another example, in the pharmaceutical industry the expiring of patents is a major metrical wave that inexorably drives: (i) the need for companies to be continually innovating, since patents on the drugs will expire in 20 years; (ii) constant mergers and acquisitions among firms. I have identified the nature and basic features of such metrical waves elsewhere (Jackson 2015) and therefore will not elaborate of them here.

And yet: the moral science of economics is much different from natural science. The discipline of economics deals with abstract objects that have been socially constructed. The abstract entities studied by economics arise from human intentionality. Accordingly, these objects have a much different ontological status from objects of nature. Is something of vital importance being left out of the picture when economic theory privileges the measurable, non-normative side of things – when it commodifies time, reducing time-value to the simple precept that "time is money" (subject of course to time-discounting calculations) to the neglect of the unmeasured (and unmeasurable) normative aspects of time-value?

More specifically, what is missing from conventional mental models of economics in their deployment of purportedly value-neutral, deterministic mathematical formulas such as the following depictions of the *time-value of money:*

As the cumulative present value of future cash flows at time t:

$$PV = \sum_{t=0}^{n} \frac{FV_t}{(1+i)^t}$$

As value diminishing over time:

$$\mathcal{L} := -\partial_t + r(t).$$

I shall argue that economic time-value is a much richer concept than the above mathematical representations of purportedly scientific temporal phenomena would suggest – a nonscientific, interpretive concept imbued with ethical and existential significance. According to my account, economic, moral and temporal values depend on and influence one another in complex ways. Intellectual responsibility about the value of time is itself an important value, and it engages concepts about which we often disagree. We disagree about economic and ethical matters in part because we carry different understandings about how time figures into what we take to be most important in our lives and the lives of others. We have different views, backed by competing justifications, about the weight to give to past, present, and future, and how best to balance these orientations across short-term, mid-range, and long-run horizons.. Moreover, we have different views about whether we should treat time as having instrumental or intrinsic value. Justifying these kinds of contested views, I maintain, involves advancing and defending with moral argumentation, interpretations about time-value and its interlinkage with other values.

Yet conventional economic theory treats the time-values accorded to economic assets, economic operations, and other economic objects[5] as if they are scientifically determinable and morally neutral empirical properties akin to the properties of objects of nature. Natural science seeks to study phenomena in nature by means of empirically verifiable and factually measurable physical properties. Students embarking on the study of economics are often instructed from the beginning that doing good science means carefully separating facts from values. The field of positive economics, they are told, is concerned with facts about the way the world actually operates. Normative economics, on the other hand, is addressed to questions of what one ought to do when se equipped with knowledge of such facts. So a broad consensus arises that good scientists will be careful to keep their moral proclivities, along with those of everyone else, at a distance from their economic analysis. To be sure, that proposition is itself a disguised normative claim. Nevertheless, upon its acceptance, everything else is taken to follow as a matter of logic.

One might suppose that the normative dimensions of time-value have already been given a satisfactory account by economic theories that have been related to empirical scientific disciplines, for instance, in Kahneman and Tversky's "prospect theory" (Kahneman and Tversky 1979) and the psychological approach taken by Zimbardo and Boyd's "time perspective theory" (Zimbardo and Boyd 2008). Don't such psychological studies about how people make choices, presumably based on their personal values and preferences, bring the human element in all of its normativity sufficiently into the mix? Yet on closer consideration, I argue, such theories function as predictive or strategic tools, in effect joining forces with the ideology of mainstream economics in evading deeper and more controversial moral issues.

The theories of Kahneman and Tversky, along with those of Zimbardo and Boyd, seek to describe the way people choose between probabilistic alternatives involving

[5] By 'economic objects' I mean abstract entities. Such thought-objects, as they occur in economic theory, purport to depict activities and events like 'equilibrium,' 'demand,' 'marginal propensity to consume', and so on.

risk, where the probabilities of outcomes are known, and describe various conventional assumptions people make in regard to time orientations, respectively, while ignoring and suppressing the contentious sphere of normative interpretive that necessarily enters into our (often unwitting) judgments about time-value.[6] There are alternative conditions under which the assumptions of "prospect theory" and "time perspective theory" either would not hold up or would fail to justify purportedly scientific claims advanced. Consider, for instance, conditions of present-centered consciousness and mindfulness pursued by contemplative practices, which arguably fundamentally alter the ethical significance of one's interpretations of past and future and associated human responses of fear, risk, desire and many others. This question is one of special interest to Buddhist economics and other spiritual and philosophical approaches to the study of economics seeking a deeper understanding of how psychology and emotions guide economic activity and the moral implications thereof.

9.2 Normative Dimension

What remains unmet is the challenge of confronting the assignment of time-value, not in its narrow purportedly empirical and descriptive guise, but in its broader evaluative and prescriptive character. This is a moral-epistemological challenge.

In an important way, such nonscientific aspects of time-value refer back to the reason why ethics itself matters. Ethics involves engaging with questions of living the good life. Economics cannot plausibly continue to ignore matters of human well-being under the pretense that it is a value-free enterprise, dictated only by objective and determined physical laws.

Although time is real,[7] I shall argue that much of what is significant about the reality of time cannot be properly grasped so long as time-value is understood merely as a matter of "plain fact." Instead, what we say about time-value, whether in a descriptive context – as in the natural sciences – or in a normative context – as in contexts such as personal ethics, social morality, law, music, and as I argue here, in economics – involves interpretation.[8] Otherwise stated, time-value in economics is best understood as an interpretive concept in contradistinction from a criterial concept or a natural-kind concept.

---

[6]As empirical research shows, if you offer someone the alternative of $100 now or $150 in 6 months, they will tend to prefer $100 now. People are inclined to prefer rewards in the present over rewards in the future. But notice that this is simply a conventional psychological tendency that does not settle the moral issue of how one *ought* to assign economic value across time orientations and time horizons. Presumably, in order to persuade people to place greater value on future rewards, the benefits associated with them need to be significantly greater than what is rationally expected.

[7]Accounting for the nature of time itself is a highly complicated and contentious undertaking, exemplified in the vast philosophical literature on the ontology and metaphysics of time.

[8]For an extended account of the idea of interpretation and interpretive concepts, see Dworkin, *Justice for Hedgehogs* (2011).

9.2.1 Time-Value as an Interpretive Concept

We may distinguish three kinds of concepts. The first type are natural-kind concepts, which are used to distinguish natural kinds such as 'tiger' and 'gold.' Such concepts "have a fixed identity in nature, such as a chemical compound or an animal species, and... people share a natural-kind concept when they use that concept to refer to the same natural kind" (Dworkin 2011, 159). The second variety of concepts are criterial. The meaning of such concepts are established on the basis of a set of defining criteria, for instance, the criteria for calling something a 'triangle' or a 'book'. For such concepts, disagreement centers around whether or not the given criteria are satisfied (Dworkin 2011, 158–159). By contrast, interpretive concepts are those for which no fixed criteria or instance-identifying decision procedure obtains. Rather one needs to turn to normative or evaluative considerations in order to come up with the best justification for whatever the total set of practices at hand happens to be. Concepts such as 'courtesy' and 'justice' are interpretive in this sense: no fixed set of criteria determines how they apply.

Why is it important to bring this distinction to bear on time-value? The reason is because the concept of time-value(s) involved in decisionmaking in areas as varied as morality, personal ethics, music, economics, and business is best understood as an *interpretive concept.* To briefly illustrate: Moral philosophers disagree about whether a future-looking consequentialist approach, or a deontological approach that includes regard for past-looking retributive factors is a better way to analyze and resolve a dilemma. Some people believe it is important to remember past wrongs inflicted by friends, while others stress the importance of forgiving such wrongs and looking to the future instead. Musicians disagree about how much rubato (discretionary giving and taking of time from strict tempo) to use in certain passages of a Brahms sonata to make the piece attain the best level of expressiveness. Economists disagree about the ethicality of high-frequency trading on the stock market. Businesspeople disagree whether firms should be focused on short-term or long-term financial performance.

Here is how Dworkin characterizes interpretive concepts in contrast to criterial and natural-kind concepts:

> Sharing an interpretive concept does depend, as sharing criterial and natural-kind concepts depends, on agreement. But the kind of agreement that is required in the case of an interpretive concept is very different: it is not agreement on a decision procedure as a decisive test for instances. On the contrary, sharing an interpretive concept is consistent with very great and entirely intractable differences of opinion about instances. It is also consistent with some people who share the concept denying that it expresses any value at all. Someone who declares that there is no value in what is expressed by a certain concept — chastity, perhaps, or etiquette or patriotism – must suppose ample agreement on paradigms of that concept among those who do count it as valuable. Otherwise his debunking argument could not take hold (Dworkin 2011, 161).

It is particularly noteworthy in this regard that Socrates' dialogues treat ethical questions as centering around criterial concepts – leading to no satisfactory answers when one queries, say "what is justice?" Aristotle, on the other hand, characterizes ethics as more like arts of medicine, navigation, comedy – revolving largely around making judgments with and about interpretive concepts. Ethics is not the same as mathematics, so asking "what is courage?" engages reasonable debate, or dialectic, about competing conceptions of courage (Hartman 2007). By contrast, to pose the question "what is the value of x in the equation '5x + 9 = 24'?" does not engage interpretive debate but rather algorithmic problem solving as dictated by mathematical logic.

A narrow and abstract context within which one confronts otherwise criterially-set ascriptions of time-value can shift to the wider real-world context in which sometimes complex clusters of apposite time-value(s) must be reckoned with, not as a matter of abstract theory, nor simply as a matter of plain fact, but instead a question of value. The narrow criterial concept then shifts to a wider interpretive context in which normative choice or moral judgment is engaged.

To illustrate time valuation occuring in wider interpretive contexts, in constitutional interpretation, the "original meaning" approach (conservative orientation) places central importance on ascertaining *past* intentions and meanings of the framers. It resists attempts to confer contemporary meanings upon the language expressing constitutional concepts. As such, there is a privileging of the value or weight accorded to past consitutional meaning. By contrast, the "living document" approach (liberal orientation) stresses a preference for keeping the constitution up-to-date with contemporary understandings of the document's language. This involves assigning greater value or weight to *present* (or "evolving" or future) constitutional meaning. Consider that in Roper v. Simmons (2005) the U.S. Supreme Court confronted the issue of whether executing a 17 year-old juvenile violates the Eighth Amendment prohibition of 'cruel and unusual punishment.' In such a legal context, there are at least two levels at which normative issues of time-value are engaged. First, there is the contest between past-privileging (conservative "original meaning" view of the kind championed by Antonin Scalia) and present-privileging or future privileging (liberal "living document" view of the sort espoused by William Brennan) approaches to constitutional interpretation. Second, (recognizing the constitutional permissibility of individual states opting for the death penalty) there is the moral choice of time-value attached to assignments of capital punishment for juveniles. Is some minimal passage of time – a person's age – a moral prerequisite for ascription of responsibility what could warrant a sentence of capital punishment?

It is helpful to turn to music for an illustration of how the idea of criterial concepts contrasts with the idea of interpretive concepts. In music time-value plays an especially vital role and is integrally connected to all other elements (Barenboim 2008). Consider the designation ♩ = MM 88. Here the time-value of a quarter note is specified to be 88 beats per minute. In this sense, the time-value is criterial. However, the time-value can shift to an interpretive concept in the following way.

Whereas in one sense the designation of a quarter note = 88 does indeed have a descriptive "plain-fact" character when set out in a musical score, nevertheless, a host of important aesthetic and even moral values come into play in any interpretations of the music that would utilize (or modify or even disregard) such a designation.

Thus, the alternative interpretations of the time-value assigned to quarter notes — which accordingly impacts the time-values of other notes and rests across an entire composition – may be disputed by various performers, conductors audiences, and critics. Should one slavishly follow such a designation even in the face of suspicions that the composer was mistaken about its performative feasibility, or evidence that his metronome was faulty? Or should one instead defer to performance tradition (Scruton 2016)? Should rubato be used, and if so, at what points and how much?

In music, a conductor or performer must decide between privileging past versus present time orientations. Should one perform with historical deference, according to the specific tempo instructions laid down by the composer? Or ought one to adjust the tempo according to other considerations of the present, such as the skill levels of an orchestra, one's own conception of the desired metric waves established by a performance (Begbie 2000), acoustics of the venue, and so on?

9.3 Efficiency Scenarios

Scenario 1 A region in the Congo may be exploited for its petroleum reserves, which is anticipated to generate enormous profits in the short-term, yet create substantial environmental damage to ecosystems. A traditional business-and-economics mindset will disregard the environmental damage since, after all, that is an externality the cost of which is to be shifted to some third party. Alternatively, the region could be used for eco-tourism, which will not be as profitable in the short-term, but will potentially offer an environmentally sustainable option. Which of the two options – petroleum exploitation, or eco-tourism – is the more efficient development path for the region of the Congo? How does so-framing this question impact the interpretations of time-value that may be brought to bear?

Scenario 2 If harvested by careful hand-picking (time-consuming, yet preserves longevity of trees) an olive crop can be gathered each year. But some farmers use a quick short-cut method, shaking the tree's branches and beating them with sticks to make olives drop. In that case, the olive crop can only be harvested every other year, and the tree does not live as long. Olive trees that have been carefully tended can be expected to live for hundreds, or in some cases, even thousands of years. There are claims of olive trees from the Garden of Gethsemane dating back to the time of Christ (Lewington and Parker 1999). Which of the olive harvesting approaches — shaking-and-beating, or hand picking – is the more efficient method? Again, how does framing the question in terms of efficiency steer the time-value inquiry at hand?

We can understand efficiency in the sense that economics typically does, as a matter of plain-fact. Efficiency is seen as something to be determined in a value-neutral, descriptive way – as a criterial concept that does not involve moral values. From this criterial vantage point, the concept of economic efficiency is a theoretical ideal.[9] It refers to a state of affairs that is presumed to be approached yet never actually attained. Typically, economics analyzes the extent of loss – called waste – obtaining between pure efficiency and reality in order to assess how efficiently an economy operates. The efficiency standard enjoys widespread acceptance in economic discourse and penetrates deep into the spheres of law and public policy. The matter attracts even more confusion given that the term efficiency carries a distinctive sense in ordinary parlance, and given that its meaning in economic policy discussions is sometimes traded upon with its meaning in the natural sciences (e.g., useful work per quantity of energy, Faraday efficiency, Quantum efficiency, Thermal efficiency, and so on), where inputs and outputs are much more easily identified and evaluated and no compensation criteria (discussed above) are called for. The widely accepted efficiency standard lends economic policy discussions a spurious aura of "scientific" credibility.

9.3.1 Opposing Viewpoint Considered

Let us consider a possible objection. A critic may say that economics should not be faulted for framing of time-value, efficiency and associated concepts as descriptive analytical tools. So long as we understand that efficiency thus conceived may be deployed in this or that way, according to other non-economic goals and social values, there is no problem. The objection will point out that economics seeks a concept of efficiency that is objective and value free. Economic outcomes will present some arrangement of goods and services, prices and situations of different individuals. In order to deal with the problem of incommensurables associated with such outcomes, and enable a given outcome to be declared more or less efficient than some other outcome, economics turns to proposed valuations that relevant individuals assign to the outcomes. Taking such an approach presupposes that individual preferences are what ought to count when reckoning issues of efficiency. However, the critic will remind us, the procedure of itself merely serves to define efficiency. Apart from providing support for the definition, no value judgment is engaged, so long as no one asserts, say, that "efficiency is good" or that "efficient outcomes are the best outcomes." Once that leap is made, one is asserting that individual preferences ought to count in determining which economic outcomes are the best. Hence, the objection might appear reasonable enough. What's wrong with defining "efficient" as "what people prefer"?

---

[9] Economic efficiency denotes an economic state wherein resources are optimally allocated so as to serve each party in the best possible way, while at the same time minimizing waste and inefficiencies (allocational efficiency). In an economy that is economically efficient, any changes taking place to benefit one party will harm another (Pareto efficiency). Concerning production, efficiency implies that goods are being produced at the lowest cost, as well as the variable inputs of production (production efficiency).

The objection, however, overlooks some crucial points. The first point is that when it comes to time, it is quality and not simply quantity that matters. The false assumption is that time can be measured, captured, controlled, and deployed as an economic variable. Economic theory treats time-value as an economic variable that, as with capital or labor, is made more effective by operations that will maximize the quantity attained within the smallest possible interval. However, this incorrectly assumes that time and productivity stand in a linear relationship that holds as a matter of plain fact. Yet because of the qualitative dimension, judgments of value, often contested and controversial, color the notion of efficiency.

A second point is that efficiency becomes distorted when it is fused to models of rational agency that are purely consequentialist or otherwise (like Parfit's approach) reductive. I will consider each of these distortions in turn.

Consequentialism advances an overall built-in temporal bias for only being concerned with future consequences in moral deliberation, yet does not provide any justification for this exclusive choice of time orientation. Some moral justifications, such as those from deontological theory or virtue ethics, may require essential reference to the past in ways that would qualify brute regard for future pleasure, happiness, preferences, or utility. Indeed, some moral justifications may advocate a magnified regard for future generations in a way that does not discount their interests, as ordinary consequentialist approaches tend to do.

How do reductive approaches distort time-value? Economics has gone off the deep end by merging its purportedly descriptive account of time-value with a peculiar reductive view of the human subject of the sort put forth by philosopher Derek Parfit (Parfit 1984) and followed by economists and behavioral scientists, among them Robert Strotz (Strotz 1955–56) and Shane Frederick (Frederick 2006).

Here is a capsulized summary of Parfit's reductive account of personal identity: At time t_1, there is a person. At a later time t_2, there is a person. These people seem to be the same person. To be sure, these people share memories and personality traits. However there are no further facts in the world that make them the same person (Parfit 1984).

Parfit wants to address the question of what makes someone the same person over time. He frames the available alternatives as including: (i) persistence of self as immaterial substance (soul); (ii) continuity of bodily self (biological animal); (iii) continuity of psychological make-up, or the thinking self. Parfit argues for a psychological continuity criterion for personal identity. What matters is the psychological connectedness of a person's memories, desires, and intentions over time.

Proceeding from the reductive view economic analysis has joined forces with behavioral psychology to propose a trisection of the person into present self, future self, and past self. Granted, some forms of reductionism would reserve room for "psychological connections" across these temporal component selves, including an individual's memories, personal characteristics and interests, which together are taken to comprise "personal identity." Thus, it has been asserted that some people have temporal selves that are well connected, meaning that one's personal identity remains stable over time. Others, according to this account, experience, or foresee, changes in their personal identity such that a discontinuity between temporal selves obtains.

Yet even with concessions made for "psychological connectedness," and while granting that the relation between personal identity and time is complex and in no way resolved, I would dissent from such temporal fragmentations of the human person on the following grounds. First, as others have argued, personal persistence is an ultimate and unanalyzable fact (Shoemaker and Swinburne 1984). Second, the self is "an enduring irreducible entity, essentially a self-conscious subject of thought and experience and source of intentional action." (Lowe 1996). Third, the human spirit and its material body forms a substantial unity; their unity constitutes a single nature (Catechism ¶365, 1994). Especially in light of the latter of these reasons, the well-known "teletransporter" thought experiment[10] that Parfit proposes fails. A person is not teletransported like that. Actually he dies.

Nevertheless the reductive position has sparked widespread fascination in the economic literature with elaborate calculations and "thought investments" about the myriad ways that such fragmented temporal selves might project and engage in time-value discounting according to their psychologically determined generic time orientation preferences. Have a look:

> Present Self mentally simulates value forward ("prospection") and a Future Self (or, to be more exact, the Present Self's projection of its Future Self) mentally simulates value backwards ("retrospection"). So, the direction of travel... is forward in time along a growth curve and back to the present along a discount curve. Prospection requires the Present Self to mentally visit the future; while retrospection requires the Projected Future Self to mentally return to the present (its past). This mental time journey begins and ends at the present time (Blake and Pickles 2014, p. 2).

The jaw-dropping oddity of these findings when deployed to calculate the time-value of money is best expressed by directly quoting a representative excerpt:

> £1 is projected forward exponentially at a growth rate of 6% to a value of £5.74 and then that future value is exponentially discounted, at a rate of 4%, to a present value of £1.77. So, £1 today is valued at £5.74 in 30 years' time; and £5.74 in 30 years' time at £1.77 today. In the second case, £1 today is projected to grow hyperbolically to £3.32 in 30 years' time and that future value is then hyperbolically discounted to a present value of £1.77. So, £1 today is valued at £3.32 in 30 years' time; and £3.32 in 30 years' time at £1.77 today. Both of these thought investments produce a present value greater than the initial value. By implication, the Hyperopic Valuer values £1 today, invested for 30 years, at £1.77 (Blake and Pickles 2014, pp. 14-15).

So the "rational agents" behind such research include the following cast of characters. There is the myopic time-valuer, who values the present due to deficiency in imagining the future or from psychological discomfort when asked to defer immediate gratification (Loewenstein and Prelec 1991, Bartels and Urminsky 2011).

[10] The teletransporter puts one to sleep, then breaks one down into atoms, copying then relaying it to Mars at the speed of light. On Mars, another machine re-creates one's body from local supplies of carbon, hydrogen, etc.) Every atom gets restored to precisely the same original arrangement. Parfit asks if teletransporter is a method of travel—if the person on Mars upon awaking is the same person as the individual entering the teletransporter back on Earth.

By contrast, the hyperopic time-valuer has negative time preference, and as such values the future over the present. This preference is in turn associated with a strong desire for accumulation, having self-control permitting resistance or abstinence from immediate gratification (Keinan and Kivetz 2008). In addition some hyperopic time-valuers are characterized as gaining pleasure from anticipation of the future and its rewards. Finally there is the neutral time-valuer who harbors no present, past, or future time bias (Greene and Sullivan 2015).

The upshot is that time-value is trapped within a brute-factual and behaviorally deterministic framework that simply takes people's supposedly hard-wired temporal preferences at face value. Such a conception derails any deeper reflective inquiry about moral reasons or justifications concerning how one ought to value time and how doing so connects to pursuit of the good life. Living well implies an attainment of temporal harmony which I maintain involves taking a certain attitude toward time and respecting moral integrity in temporal judgments. The attitude I have in mind is seeing time as an existential gift that is both valuable and at the root of how we assign value to all things. Moral integrity involves taking responsibility for making one's own time-sprung moral judgments and accepting accountability for acting consistently on them over time.

A third point is that many difficulties surround determining what "people prefer." As well it is far from obvious what such a determination even means. In circumstances where all affected individuals will evidently gain (it's clear they all prefer some change being made) then a Pareto improvement is pronounced. It gets stickier, however, if some individuals will gain yet others will lose. For these sorts of mixed scenarios economics looks to compensation tests to ascertain whether or not the gains will offset the losses. By making this move one is essentially saying that so far as efficiency is concerned, a distribution of gains between individuals is either irrelevant or a matter about which economists ought to remain neutral. A typical response is that while the distribution may be relevant, it constitutes a separate matter. One should first make the pie as big as possible before thinking about how to slice it up. Making that move introduces a value judgment. It introduces a less defensible case than the simple case in which no one loses. Such a position suggests a sort of democratic approach whereby the majority of individuals prefer outcome x, or the minority of losers cannot prevail against the winners with regard to changes that are clearly beneficial. Other difficulties concern the question of non-uniqueness and potentially infinite number of Pareto optima. No distinctly economic approach would justify a choice among the optima but would need to be driven by reference to noneconomic values. In addition, it is alleged by some that accepting the reasoning of Pareto logic is already inclined in favor of a present state of affairs – a "tyranny of the status quo" (Acocella 1998, 32) – thus introducing a temporal value judgment.

A fourth point, related to the one made above about the problem of determining individual preferences, concerns the deployment of hypothetical market valuations and of postulating unknowable counterfactuals for unknowable and/or unimaginable future states.

A closely related fifth point is that the problem extends further than the conventional efficiency standard as such into its application to the static model of resource allocation. Static efficiency criteria are not relevant in a dynamic context in which market processes are occurring in real time. Such processes are fueled by a multiplicity of perceptions and expectations that people have concerning the value of resources, as well as of the very processes themselves, which stay open-ended instead of in equilibrium. Since novel technologies depend on future knowledge not available in the present, one is not able to include future technologies within a present-situated choice set. Nor can one include the novel products, production methods, and organizational modes that new technologies will bring about. Change in technology arises from a complex interplay of individual ideas that remain incomplete and indeterminate.

Suppose after all of this, the critic remains unpersuaded, insisting that economics can simply readjust the ideas of economic value and efficiency so as to render them more "time sensitive." But the critic has failed to grasp the heart of the problem. It is not merely that scientifically specified criterial concepts must be spelled out in greater multiplicity as time sensitive phenomena amenable to empirical description. In addition it is necessary to expose the myth of time as a value-neutral component of economic theorizing. My argument is that we better understand efficiency, not in the criterial factual sense that mainstream economics offers, but instead as an interpretive concept, one that involves making moral judgment. Ultimately, the moral judgment applies to time-value.

9.4 An Interpretive Standpoint

An interpretive standpoint is a place from which one views the world. Such a standpoint guides what one focuses on and what one values. As well, an interpretive standpoint influences what goes unnoticed and what is not valued. As opposed to the traditional time-as-plain-fact presupposition or conceit that underwrites the conception of efficiency used by positivist economics, a nonpositivist interpretive standpoint on economics includes normative interpretations beyond merely descriptive accounts of temporal economic phenomena. That is, from a nonpositivist interpretive standpoint, economics encompasses both facts about time and norms about time-value.

- A standpoint influences how people adopting it socially construct the field of economics.
- A standpoint constitutes a mental model from which economic scenarios are viewed.
- A standpoint is a position from which economic objects and economic principles are viewed and according to which they are compared and judged.

A deliberator confronts alternatives from which to choose. What gives the deliberator a justified interpretive standpoint?

From some standpoint, the deliberator will render an assertion with interpretive content I_1.

A collection of statements C is made up of statements S.

The statements are arranged in some reasonable manner R_1 which specifies the content of the norms and facts. These are in turn connected to appraisals A_1 of time (T) and value (V). The appraisals of time range across the criteria (past/present/future) of time-orientation (O), time-horizon (H) criteria (short, mid, and long-term) and contain weightings (W) in a serial ordering to which rankings can be assigned, from 1, 2, 3, etc. The judgments of value may range across the spectrum of economic values (such as efficiency, sustainability: $E_{e,s}$), aesthetic values (such as beauty and interest: $A_{b,i}$), or moral values (such as goodness and justice: $M_{g,j}$).

$$I_1 : C_1 R_1 \left(S_{1\ldots} S_n\right) \Psi A_1 [T\left(O_{pa,pr,fu}\right) \wedge \left(H_{st,mt,lt)} \wedge \left(W_{\underset{a}{1,n}}\right) \wedge V\left(E_{e,s} \quad A_{b,i} \quad M_{g,j}\right)\right]$$

Not only do our values penetrate our understanding of economic time-value, but also, time penetrates our values. Our day-to-day use of seemingly "descriptive" units in dealing with the time in science are ultimately based on a conventionally chosen metric framework of our own creation: second, minute, year, and so on. The fact that we use fixed durational units itself reveals the imposition of peculiar values: consistency, cyclicality, treating time as having a uniform character in the way we take space to be.

Conversely, time penetrates into the very heart of our values. That is, the very idea of value – and our ascriptions of value – both depend upon a temporal context. One appreciates something for a duration; money is valuable as a means of exchange, but exchange presupposes a background time context within which exchanges are possible.

The prices of shares, commodities, and futures are, in a common sense view, necessarily connected to, and are indeed thought to be relative to a particular point in time (albeit sometimes only a millisecond, in the case of high frequency trading[11]) within the operation of the stock market.

Many mainstream economists consider the time element figuring into economic efficiency to be something value-neutral and scientifically determinate and objective. Accordingly, the supposedly value-neutral time of economic efficiency is detached from the idea of human temporality – and attendant values (aesthetic, moral, spiritual).

[11] High Frequency Trading (HFT) involves the execution of complicated, algorithmic-based trades by powerful computers. The aim of HFT is to take advantage of minute discrepancies in prices and trade on them quickly and in huge quantities (MacKenzie 2011).

9.5 Conclusion

Modern economics deploys the concept of time-value as if it were a descriptive, empirically-verifiable objective property of the sort studied by natural science in its investigation of the natural world. In particular, economics frames efficiency as a core feature of time-value understood in a purportedly descriptive sense. Economic time-value is stripped of its normative quality, and held up for scientific investigation and brute factual description of the sort associated with criterial and natural-kind concepts.

As such, economics deploys the conceptual devices such as the "time-value of money" in its quest to objectively ascertain in a purportedly morally-neutral fashion behaviors that are favorable to capital efficiency maximization.[12] Yet, as I have shown, economics thus removes the relation of efficiency to time-value from the realm of normative choice – that is, from the realm of contested interpretive concepts.

References

Acocella, N. 1998. *The foundations of economic policy: Values and techniques*. Cambridge: Cambridge University Press.

Barenboim, D. 2008. *Everything is connected: The power of music*. London: Weidenfeld and Nicolson.

Bartels, D., and O. Urminsky. 2011. On intertemporal selfishness: The perceived instability of identity underlies impatient consumption. *Journal of Consumer Research* 38 (1): 182–198.

Begbie, J.S. 2000. *Theology, music and time*. Cambridge: Cambridge University Press.

Blake, D., and J. Pickles. 2014. *Value-Time Curve Psychology*. Discussion paper PI 1403. http://www.pensions-institute.org

Catechism of the Catholic Church. 1994. Doubleday. New York.

Davis, M., A. Kumiega, and B. Van Vliet. 2012. Ethics, finance, and automation: A preliminary survey of problems in high frequency trading. *Science and Engineering Ethics*. https://doi.org/10.1007/s11948-012-9412-5.

Dworkin, R. 2011. *Justice for hedgehogs*. Cambridge: Harvard University Press.

Frederick, S. 2006. Valuing future life and future lives: A framework for understanding discounting. *Journal of Economic Psychology* 27 (1): 667–680.

Greene, P., and M. Sullivan. 2015. Against Time Bias. *Ethics* 125 (4):947–970.

Hartman, E.M. 2007. Socratic questions and Aristotelian answers: A virtue-based approach to business ethics. *Journal of Business Ethics* 78 (3): 313–328.

Hutton, W., and A. Jones. 2005. Time and money. In *About time: Speed, society, people and the environment*, ed. T. Aldrich, 88–105. Sheffield: Greenleaf Publishing.

Jackson, K.T. 2015. Aesthetics, human rights and economic life: Temporal perspectives. In *Business and the greater good: Rethinking business ethics in an age of crisis*, ed. K.J. Ims and L.J.T. Pedersen. Cheltenham: Edward Elgar.

[12] In economics, return on capital is a profitability ratio that measures the return that a given investment generates for capital contributors such as stockholders and bondholders. Return on capital is thus taken as a measure of how effective a company is at turning capital into profits.

Kahneman, D., and A. Tversky. 1979. Prospect theory: An analysis of decision under risk. *Econometrica* 47 (2): 263–291.
Keinan, A., and R. Kivetz. 2008. Remedying hyperopia: The effects of self-control regret on consumer behavior. *Journal of Marketing Research* 46 (6): 676–689.
Lewington, A., and E. Parker. 1999. *Ancient Trees*. London: Collins and Brown.
Loewenstein, G., and D. Prelec. 1991. Negative time preference. *American Economic Review* 81 (3): 347–352.
Lowe, E.J. 1996. *Subjects of experience*. Cambridge: Cambridge University Press.
MacKenzie, D. 2011. How to make money in microseconds. *London Review of Books* 33 (10): 16–18.
Parfit, D. 1984. *Reasons and persons*. Oxford: Clarendon Press.
Scruton, R. 2016. *Understanding music: Philosophy and interpretation*. London: Bloomsbury Academic.
Shoemaker, S., and R. Swinburne. 1984. *Personal identity*. Oxford: Blackwell.
Strotz, R.H. 1955-56. Myopia and inconsistency in dynamic utility maximization. *Review of Economic Studies* 23 (3): 165–180.
Zimbardo, P., and J. Boyd. 2008. *The time paradox: The new psychology of time that will change your life*. New York: Simon and Schuster.

Chapter 10
The Crucial Role of Reactivity in Economic Science

Bruno S. Frey

Abstract Many academic economists take it as a matter of course that economics should become a natural science. Such a characterization misses an essential aspect of a social science, namely reactivity, i.e. human beings systematically respond to economic data, and in particular to interventions by economic policy, in a foreseeable way. To illustrate this finding, the paper uses examples from different fields: happiness policy, world heritage policy, and science policy. Reactivity requires a different policy approach from governments claiming to maximize social welfare in the form of happiness. What is needed are basic constitutional provisions, or "rules of the game", decided behind the veil of ignorance. Moral considerations must enter when the basic rules setting the way in which individuals interact are determined. Moral aspects are also indispensable as guidance to individuals when they take decisions within these rules. It is most unlikely that intelligent learning machines can fulfill this task in an acceptable manner.

There is a fundamental difference between the natural and the social sciences due to reactivity. This difference remains even in the age of Artificially Intelligent Learning Machines and Big Data. Many academic economists take it as a matter of course that economics should become a natural science. Such a characterization misses an essential aspect of a social science, namely *reactivity*, i.e. human beings systematically respond to economic data, and in particular to interventions by economic policy, in a foreseeable way. To illustrate this finding, I use three examples from quite different fields: Happiness policy, World Heritage policy, and Science policy.

Thanks are due to Margit Osterloh, Arjo Klamer, Armin Steuernagel, Christian Ulbrich, Jonas Friedrich, Peter Róna, and László Zsolnai for helpful remarks.

B. S. Frey (✉)
University of Basel, Basel, Switzerland

CREMA–Center for Research in Economics, Management and the Arts, Zurich, Switzerland
e-mail: bruno.frey@bsfrey.ch

© Springer International Publishing AG, part of Springer Nature 2018

P. Róna, L. Zsolnai (eds.), *Economic Objects and the Objects of Economics*,
Virtues and Economics 3, https://doi.org/10.1007/978-3-319-94529-3_10

10.1 Economics as a Natural Science

Many academic economists, among them very famous ones, take it as a matter of course that economics should become a natural science. It is *the* great example to be followed as strictly as possible with respect to rigor, model orientation and the use of mathematics.[1]

This view has a long history, as exemplified in the "Scoping Paper for the Symposium concerning Economic Objects and the Objects of Economics" held Oxford in July 2017. Thus, in 1792 Condorcet explicitly claims that the social sciences should imitate the physical sciences. So does Edgeworth in 1881. That this still holds today is shown by Gregory Mankiw (2018: 20) who in his leading textbook *Principles of Economics* states:

> " Economists … approach the subject of the economy in much the same way as physicists approach the study of matter … they devise theories, collect data and then analyze these data in an attempt to verify their theories".

I will argue in this contribution that such a characterization misses an essential aspect of a social science, namely *reactivity,* i.e. that human beings, acting alone or in groups, systematically respond to economic data, and in particular to interventions by economic policy in a foreseeable way. Individuals do not passively accept what economists suggest on the basis of economic theory and econometric analysis. Rather, they react and therewith transform the data. It follows that "data" are not just "given" but they will be subject to the manipulation by human beings. (The term "data" is indeed a misnomer; it is the plural of "datum", the Latin word for "given". As people react and transform and manipulate the data, they are not given). This means that – unlike in the natural sciences – it is not (generally) possible to make reliable predictions. Even with the advent of Big Data, economic forecasts do not improve in precision but, if anything, get worse.

The crucial importance of reactivity for economic policy is discussed using three examples from three quite different fields: Happiness policy, World Heritage policy, and Science policy. These examples are generalized, pointing out that the reactivity effects may have quite different consequences for society.

The insights gained suggest a much different approach to economic policy, which will be shortly sketched.

[1] There are, of course exceptions. A notable and most recent one is by Ariel Rubinstein (2017), himself a leading and highly mathematical game theoretician. He strongly attacks the over-mathematics in economics both with respect to educating students and undertaking research. It should be immediately added that such a critique has a chance to be published in a top academic journal only by a mathematical scholar, i.e. outsiders would be considered to be incompetent to judge.

10.2 Reactivity: Economics as a Social Science

Reactivity is a fundamental feature of any social science (see e.g. Espeland and Sauder 2007) including economics; it has in particular been noted in finance (e.g. Soros 2013 who uses the term "reflexivity"). Individuals are not passive human beings accepting economic, political or social conditions, and in particular they react to interventions coming from outside, such as by governmental policy. This insight is, of course, not new at all. After all, game theory has been developed to explicitly take into account that people react to each others' actions and to other external influences. Even a "stable" equilibrium among actors shifts if exogenous conditions change, for instance, if there is a natural disaster or a terrorist attack not taken into account in the model.

Surprisingly enough, reactivity is largely disregarded even in modern economics though game theory has partly been developed by an economist (Oscar Morgenstern) and is part of the curriculum every economics student has to learn. But as the citation above from Mankiw's (2018) textbook shows, reactivity is largely disregarded. Indeed, Mankiw speaks of the (passive) "matter" rather than of (reacting) *human beings* one has to study.

The difference between the natural sciences and the social sciences can be well demonstrated for the case of predictions. When a forecast is made, nature does not react but human beings often do.[2] Thus, when meteorologists *correctly* predict[3] that there will be a heavy and destructive storm, nature does not *therefore* change its course. In contrast, people have an incentive to change their behavior so that the effects of the heavy storm will be mitigated or do not occur at all. The prediction of a destructive effect then does not occur. In this sense, the prediction was self-destructing. There are also self-fulfilling predictions, for example when somebody indicates that a bank will have problems meeting its obligations in the future. Even if that prediction as such is unfounded, it tends to induce people to withdraw their holdings with the bank, which in turn may lead the bank to not being able to fulfill its obligations (see Venn 1888; Morgenstern 1928; Merton 1936). The bankruptcy of that bank may lead to a general banking crisis and to an economic downturn. Such a development demonstrates that in the social sciences reactivity must be an essential element of the analysis.

---

[2]The emphasis is on the reactions by human beings. This is different from the observer effect in physics proposing that observing a situation or phenomenon necessarily changes that phenomenon. For example, it is impossible to see an object without light hitting the object, and causing it to emit light. The object experiences a change.

[3]Meteorologists do not always correctly predict the weather because this is extremely difficult to do. Moreover, at least local weather forecasters are sometimes pressured by interest groups (tourist and sports industry) to predict better weather than they objectively believe to be true in order not to prevent people to visit leisure resorts and sports events. It may even be that the weather is influenced when a lot of people are induced to use their cars and therewith pollute the environment. The same view is expressed by Róna (2017).

The adherents of the Big Data movement are convinced that predictions will generally greatly improve with the huge amount of data now available (in particular due to the Internet of Things, see e.g. Pentland 2014). In a joint study with Reto Cueni (Cueni and Frey 2014) we empirically show that this claim is mistaken: Using more and more data only improves the forecasting quality of natural phenomena but not the quality of forecasts dealing with human action and reactivity. We compare the predictions of the weather in six different Swiss cities with the predictions made by analysts about profits of companies represented at the two largest American stock exchanges (NYSE and NASDAQ) as well as to predictions of art experts for the selling prices of Chinese art at auctions. Each data set covers the period of 12 years, from 1999 to 2010. The precision of the predictions is measured by the yearly median of the absolute prediction error as a per cent of the actual outcome. We find that the forecasting error for the weather continually (and in a statistically significant way) falls over time. The prediction of this natural phenomenon improves due to more and better data as well as to better forecasting models. On the other hand, the predictions of future firm profits and auctions prices for Chinese art do not improve over time. Indeed, if anything, the forecasting errors tend to increase over time (though not in a statistically significant way). The reason is that people react to the forecast. If prospective buyers of Chinese art believe that the future auction prices will rise to the extent predicted, they attempt to already buy now, or turn to other art. Such reactivity has a significant effect on prices, often totally changing the outcome originally predicted. Indeed, it is the purpose of forecasts of economic downfalls to muster reactions such that the unfortunate outcome does not come true.

10.3 Relevance of Reactivity for Policy

The dependence of the social sciences on the reactivity of human beings is of great importance and should not be considered an irrelevant difference to the natural sciences. This will be illustrated with three policy areas: Happiness policy, World Heritage policy, and Science policy.

10.3.1 Happiness Policy

Modern happiness research has been jointly developed by psychologists and economists (for surveys see e.g. Kahneman et al. 1999; Lane 2000; Easterlin 2002; Frey and Stutzer 2002; Layard 2005; Frey 2008; Stutzer and Frey 2010; Frey et al. 2014; Helliwell 2015; Helliwell et al. 2017; Clark et al. 2016; Ims 2017). It differs basically from happiness research undertaken over centuries by philosophers by its empirical orientation. The most important measure of happiness is subjective life satisfaction where a representative group of persons are asked: "Taken overall, how satisfied are you with the life you lead?" Careful research has established that the answers to this question are quite reliable as they correlate systematically with features normally attributed to happiness, such as being open, social and optimistic, having few problems at the work place, being less in psychic care, and committing less suicide attempts. The results of these surveys are analyzed with advanced econometric methods, also taking into account effects of reverse causation.[4]

Some of the insights of empirical happiness research correspond to common sense expectations, such as that higher income, more favorable personal relationships, better health, and living in a democratic and decentralized polity raises subjective well-being.[5] Other results are more surprising, for instance that the self-employed – who work longer hours, have to carry higher risk, and on average get a lower income – are happier. According to standard economic theory more money and leisure always contribute to utility. Empirical studies on well-being suggest that the opposite also holds. Those giving money to other persons, or engaging in voluntary unpaid work are more satisfied with their lives than those persons giving less or nothing to others. Many people also believe that artists must be unhappy in order to have an impetus to be creative. Empirical research suggests that this is not the case. Artists are happier than other persons mainly due to the greater autonomy in work they enjoy (Steiner 2017).

The happiness, or subjective well-being, scores collected from individuals may be aggregated to the country level. This allows us to compare countries according to the (average) happiness level of their population. Accordingly, the UN yearly publishes a "Ranking of Happiness". The newest "World Happiness Report 2017" lists 155 countries. The happiest are Norway, Denmark, Iceland and Switzerland; the unhappiest Syria, Tanzania, Burundi and the Central African Republic. The United States are ranked 14, Germany 16, the United Kingdom 19, France 31, Spain 34, and Italy 48.

The possibility to measure happiness as a proxy for social welfare has led a number of leading scholars in the field (such as Lord Richard Layard, John Helliwell, or Jeffrey Sachs) to propose that governments should maximize the happiness of the population. This idea fulfils an old dream in economics of "quantitative economic policy" as envisaged by Theil (1964) and Tinbergen (1956). Several countries follow this advice. Best known is Bhutan (who did so considerable time ago), France, Germany and the United Kingdom. The idea is that governments should undertake policies on the basis of the insights of happiness research, for instance with respect to health, education, culture, or sports.

Such an approach disregards reactivity. There are two reactions, which undermine the idea that governments should maximize happiness (Frey and Stutzer 2012; Stutzer 2017). First, the persons asked about their subjective happiness have a strong incentive to respond strategically. In particular, if a respondent is a strong

[4] For instance, persons in good health are happier, but happy people are less prone to fall prey to contagious diseases.

[5] Though there are still people who believe that persons living in developing countries are happier than those in developed ones, presumably because they are thought to be less affected by stress. All serious empirical evidence rejects this view.

left-winger and the government is right wing, he or she has a tendency to report a lower level of happiness than in reality in order not to honor the despised government politicians. The reverse is likely to happen for a right-wing person under a left-wing government. Second, the politicians in power have an incentive to manipulate the aggregate happiness indicator in their favor. Everyone knowing some statistics is well aware that this is quite easily possible.

As a result, the results of the happiness surveys no longer reflect "reality" and are no longer dependable as a guide to policy. The reactivity induced by the goal of maximizing an aggregate social well-being indicator does not only distort reality but even provides politicians with new instruments to deceive citizens and to remain in power.

10.3.2 World Heritage Policy

The UNESCO is engaged in the laudable goal to protect the common heritage of mankind. There is now a list of approximately thousand sites, mostly monuments, which should be protected against demolition and damage (see e.g. Frey and Steiner 2011). The sites included in the list are publicized both by UNESCO and the media, and they bear a sign officially indicating this honor.

Again, reactivity has not been taken into account. Unfortunately, in several wars and internal rebellions these sites have been attacked and partly or even wholly demolished *exactly because* they are part of the officially acknowledged heritage of mankind. Thus, in the Balkan War, the city of Dubrovnik was shelled in order to impose the highest possible damage to the adversary. The so-called "Islamic State" undertook a similar action in Palmyra. In 2001, the Taliban destroyed the world's tallest Buddha statues, which towered over Afghanistan's Bamiyan Valley for centuries. Thus, totally against the intention of UNESCO, reactivity has in this respect led to extremely bad consequences.

10.3.3 Science Policy

Science policy today strongly relies on rankings based on publications and citations. Individual scholars, universities and other research institutions are ranked according to this measure.[6] This is a rather new development. In former times nobody would have thought to rank Newton, Kant or Einstein on the basis of such measures. They were considered great scholars because of their contributions to science, which were identified in term of *content*, not of the number of publications or citations.

---

[6]These measures are most doubtful and should, in fact, not be used. See e.g. the International Mathematical Union (2014), or the extensive discussion in Osterloh and Frey (2015).

Here again reactivity undermines this policy. Individual scholars adapt by publishing as much as possible, and by driving up the number of citations by implicit or explicit cartels with colleagues. At the level of scientific institutions reactivity is also very strong because the amount of money received by the state and other sponsors depends on the rank. King Saud University improved its rankings by *several* hundred positions by offering a honorarium of 50,000 Euros to prolific and famous scholars if they indicate Saud University when publishing (Bhattacharjee 2011). In China a fully developed market for authorship in renowned scientific publication emerged. For 10,000 Euros one can claim main authorship in such a journal (Hvistendahl 2013).

10.4 Generalizing the Reactivity Effect

In each of the three policy areas here discussed – happiness, world heritage and science policy – reactivity plays a major role, setting the social sciences apart from the natural sciences, and leading to unfortunate outcomes. There are, of course, many other areas in which reactivity works in an unexpected direction, for instance in the form of crowding-out effects (Frey 1997).

There is a serious *counterargument* to this view: The social sciences must take reactivity into account in a calculated way. In that case the difference between the natural and social sciences disappears because the reactions of individuals and organizations to an intervention are included in the model. It is claimed that this procedure is enabled by the great progress made using an "Ultimate Learning Machine" built on a "Master Algorithm" (Domingos 2015). It would also allow producing precise forecasts with Big Data.

A valiant attempt to take into account reactivity has been undertaken by entertaining the idea of "rational expectations" (Muth 1961; Lucas 1976). However, the concept today has been strongly questioned (Blinder 1998; Frydman and Phelps 2013). Most importantly, rational expectations theory does not, and cannot, include human creativity being able to react in new and unexpected ways.

I consider the idea to take reactivity into account in a calculated way to be unreachable – at least in the foreseeable future (which we have to deal with). There are three major reasons why the counterargument does not hold:

1. A game theoretic model would be required which takes into account all reactions by all relevant actors. Even if such a huge and complete model could be constructed, it remains open if an equilibrium exists and if it is stable, i.e. whether there is a situation in which no actor has an incentive to deviate from his or her individually optimal strategy.
2. A model meeting the requirements of the natural sciences must be able to account for the many different perceptions people form when confronted with an intervention. Such interpretations are important for predicting the reactions, but they

are difficult or impossible to anticipate because they may be formed due to very specific conditions and states of mind.

3. It seems impossible to foreseen all kinds of reactivity. Human beings often exhibit immense creativity to react to outside interventions. Innovative reactions are by definition not predictable – not even by a "Master Algorithm" – because they would otherwise have already been used. Whether a learning machine can foresee such creative reactions boils down to how innovative human beings are considered to be. In my view they are creative and able to find completely new ways to react. If this were the case, not even a very extensive game theoretic model would suffice.

The effort to have an "Ultimate Learning Machine" to completely solve social issues in the same (mechanistic) way as done in natural science paradoxically leads to a stronger emphasis on human behavior and social relationships. If the world of learning algorithms wants to make progress in the direction of the natural science, reactivity must be fully understood: the type of reaction, its direction, strength and speed. I argue that this is an impossible task for the problems with which we are currently faced.

A totally encompassing model of society seems impossible. Rather, this idea follows the lead of technocratic social planners whose complete error was dramatically exposed by the breakdown of the socialist planning models in the Soviet Union and elsewhere. Technocrats (such as Domingos 2015) have to believe in the almighty power of algorithms. They will certainly much change our societies but it is most doubtful whether they can adequately deal with creative reactivity.

10.5 Conclusions

I propose that there is a fundamental difference between the natural and the social sciences due to reactivity. I submit that this difference remains even in the age of Artificially Intelligent Learning Machines and Big Data.

This view requires a different policy approach from governments claiming to maximize social welfare in the form of happiness. What is needed are basic constitutional provisions, or "rules of the game", decided behind the veil of ignorance (as suggested by Buchanan and Tullock 1962; Brennan and Buchanan 1985; Mueller 1996). Economic researchers, in particular those dealing with happiness, can inform the citizens about the relationships they have identified. On that basis, they may propose policies in which both their advantages and disadvantages are carefully communicated. It must then be left to the individuals as citizens how they evaluate these options. It seems impossible that a machine implanted in our brains is able to perform that task – and if so, we are in a world of machines and not of human beings.

What is the role of moral science? It is indispensable on two levels. Moral considerations must enter when the basic rules setting the way in which individuals interact are determined. Moral aspects are also indispensable as guidance to individuals when they take decisions within these rules. Again, it is most unlikely that intelligent learning machines can fulfill this task in an acceptable manner.

References

Bhattacharjee, Y. 2011. Saudi universities offer cash in exchange for academic prestige. *Science* 334 (61): 1344–1345.

Blinder, A.S. 1998. *Central banking in theory and practice*. Cambridge, MA: MIT Press.

Brennan, G., and J.M. Buchanan. 1985. *The reason of rules*. Liberty Fund: Constitutional Political Economy. Indianapolis.

Buchanan, J.M., and G. Tullock. 1962. *The Calculus of consent*. Logical Foundations of Constitutional Democracy. Ann Arbor: University of Michigan Press.

Clark, A., S. Flèche, R. Layard, N. Powdthavee, and G. Ward. 2016, December 12. Origins of happiness: Evidence and policy implications. *VOX CEPR's Policy Portal*.

Cueni, R., and B.S. Frey. 2014. Forecasts and reactivity. *CREMA Working Paper, Center for Research in Economics, Management and the Arts*, No 10–2014.

Domingos, P. 2015. *The master algorithm. How the quest for the ultimate learning machine will remake our world*. London: Penguin.

Easterlin, R.A., ed. 2002. *Happiness in economics*. Cheltenham: Edward Elgar.

Edgeworth, F.Y. 1881. *Mathematical psychics: An essay on the application of mathematics to the moral sciences*. London: C. Kegan Paul & Co. Reprinted New York: A.M. Kelley, 1967.

Espeland, W.N., and M. Sauder. 2007. Rankings and reactivity: How public measures recreate social worlds. *American Journal of Sociology* 113 (1): 1–40.

Frey, B.S. 1997. *Not just for the money*. Cheltenham/Brookfield: Elgar.

———. 2008. *Happiness: A revolution in economics*. Cambridge, MA: MIT Press.

Frey, B.S., and L. Steiner. 2011. World heritage list: Does it make sense? *International Journal of Cultural Policy* 17 (5): 555–573.

Frey, B.S., and A. Stutzer. 2002. *Happiness and economics: How the economy and institutions affect Well-being*. Princeton/Oxford: Princeton University Press.

———. 2012. The use of happiness research for public policy. *Social Choice and Welfare* 38 (4): 659–674.

Frey, B.S., J. Gallus, and L. Steiner. 2014. Open issues in happiness research. *International Journal of Economics* 61 (2): 115–126.

Frydman, R., and E.S. Phelps. 2013. Which way forward for macroeconomics and policy analysis? In *Rethinking expectations. The way forward for macroeconomics*, ed. R. Frydman and E.S. Phelps, 1–45. Princeton: Princeton University Press.

Helliwell, J.F. 2015. The economics of happiness. *NBER Reporter* 14 (2).

Helliwell, J.F., R. Layard, and J. Sachs. 2017. *World happiness report*. New York: Sustainable Development Solutions Network.

Hvistendahl, M. 2013. China's publication bazaar. *Science* 342 (6162): 1035–1039.

Ims, K.J. 2017. Happiness and human flourishing. In *Economics as a moral science*, ed. P. Róna and L. Zsolnai, 79–94. Cham: Springer.

International Mathematical Union. 2014. *Recommendation on the evaluation of individual researchers in the mathematical sciences*. Endorsed by the IMU General Assembly on August 10, 2014.

Kahneman, D., E. Diener, and N. Schwarz, eds. 1999. *Well-being: The foundations of hedonic psychology*. New York: Russell Sage Foundation.

Lane, R.E. 2000. *The loss of happiness in market economies*. New Haven/London: Yale University Press.

Layard, R. 2005. *Happiness: Lessons from a new science*. New York: Penguin Press.
Lucas, R.E., Jr. 1976. Econometric policy evaluation: A critique. In *The Phillips curve and labor markets*, ed. K. Brunner and A.H. Meltzer, 19–46. New York: Elsevier.
Mankiw, N.G. 2018. *Principles of economics*. Boston: Cengage Learning.
Merton, R.K. 1936. The unanticipated consequences of purposive social action. *American Sociological Review* 1 (6): 894–904.
Morgenstern, O. 1928. *Wirtschaftsprognose: Eine Untersuchung ihrer Voraussetzungen und Möglichkeiten*. Berlin: Springer.
Mueller, D.C. 1996. *Constitutional democracy*. New York: Cambridge University Press.
Muth, J.F. 1961. Rational expectations and the theory of price movement. *Econometrica* 29 (3): 315–335.
Osterloh, M., and B.S. Frey. 2015. Ranking games. *Evaluation Review* 39 (1): 102–129.
Pentland, A. 2014. *Social physics*. New York: Penguin.
Róna, P. 2017. *Objects of nature and objects of thought*. Oxford University: Mimeo/Blackfriars Hall.
Rubinstein, A. 2017, March. Comments on economic models, economics and economists: Remarks on economics rules by Dani Rodrik. *Journal of Economic Literature* 55 (1): 162–172.
Soros, G. 2013. Fallibility, reflexivity, and the human uncertainty principle. *Journal of Economic Methodology* 20 (4): 309–329.
Steiner, L. 2017. Artists are poor and unhappy. In *Economic ideas you should forget*, ed. B.S. Frey and D. Iselin, 135–136. Cham: Springer.
Stutzer, A. 2017. Governments should maximize the happiness of the population. In *Economic ideas you should forget*, ed. B.S. Frey and D. Iselin, 145–146. Cham: Springer.
Stutzer, A., and B.S. Frey. 2010. Recent advances in the economics of individual subjective well-being. *Social Research* 77 (2): 679–714.
Theil, H. 1964. *Optimal decisions for government and industry*. Amsterdam: North Holland.
Tinbergen, J. 1956. *Economic policy: Principles and design*. Amsterdam: North Holland.
Venn, J. 1888. *The logic of chance: An essay on the foundations and province of the theory of probability, with especial reference to its logical bearings and its application to moral and social science, and to statistics*. London: Macmillan.

Chapter 11
Economic Actors and the Ultimate Goal of the Economy

László Zsolnai

Abstract Mainstream economics employs a rather simplified picture of economic systems. Economic actors are grouped into three categories, namely individuals/households, firms, and the state. Among these actors only monetized transactions are considered. The ultimate goal of the economy is defined as maximization of individual income or financial wealth. In the reality of the twenty-first century the working of the economy is far more complex. Economic actors represent diverse types and the monetized transactions are only one from of the interactions among them. The paper suggests that we should consider economic actors such as individuals, organizations, communities, networks, and robots (or algorithms). Among these actors four basic types of interactions can be identified: market exchange, bureaucratic coordination, ethical transaction, and aggressive action. Different types of interactions may be combined but "crowding in" and "crowding out" effects may happen among them. For example, market exchange may crowd out ethical coordination. The ultimate goal of the economic system is not to produce material gains but to provide livelihood for those who are involved. Just and sustainable prosperity can be considered as the ultimate goal of the economy against which the performance of any economic system should be measured.

Mainstream economics employs a rather simplified picture of economic systems. Economic actors are grouped into three categories, namely individuals/households, firms, and the state. Among these actors only monetized transactions are considered. The ultimate goal of the economy is defined as maximization of individual income or financial wealth.

In the reality of the twenty-first century the working of the economy is far more complex. Economic actors represent diverse types and the monetized transactions are only one form of the interactions among them. The paper suggests that we should minimally consider the following types of economic actors: individuals, organizations, communities, networks, and robots (or algorithms).

L. Zsolnai (✉)
Business Ethics Center, Corvinus University of Budapest, Budapest, Hungary
e-mail: zsolnai@uni-corvinus.hu

© Springer International Publishing AG, part of Springer Nature 2018

P. Róna, L. Zsolnai (eds.), *Economic Objects and the Objects of Economics*, Virtues and Economics 3, https://doi.org/10.1007/978-3-319-94529-3_11

Among economic actors at least four basic types of interactions can be identified: market exchange, bureaucratic coordination, ethical transaction, and aggressive action. Different types of interactions may be combined but "crowding in" and "crowding out" effects may happen among them. For example, market exchange may crowd out ethical transaction or bureaucratic coordination may crowd out market exchange.

The ultimate goal of the economic system is not to produce material gains but to provide livelihood for those who are involved. Just and sustainable prosperity can be considered as the ultimate goal of the economy against which the performance of any economic system should be measured.

11.1 Physics Envy

The economics profession is often characterized by the so-called "physics envy". Economists usually admire physics for its exactness, and high explanatory and predictive power. I think that there are good reasons for physics envy. One reason is that physics (especially quantum physics) employs a rich and verified ontology.

The Standard Model of quantum physics (Mann 2010) classifies all known elementary particles and describes three of the four known fundamental forces in the universe, the electromagnetic, weak, and strong interactions (but does not include gravitational force). Although the model leaves some phenomena unexplained and falls short of being a complete theory of fundamental interactions, it has huge explanatory and predictive power.

According to the standard model of quantum physics elementary particles look like this. There are six quarks (up, down, charm, strange, top, bottom), and six leptons (electron, electron neutrino, muon, muon neutrino, tau, tau neutrino). Gauge bosons are force carriers that mediate the strong, weak, and electromagnetic interactions. The Higgs boson plays a unique role as it generates the masses of the leptons (electron, muon, and tau) and quarks. (Mann 2010)

In contrast with quantum physics, mainstream economics employs a rather simplified ontology. Economic actors are grouped into three categories, namely individuals/households, firms, and the state. Among these actors only monetized transactions are considered. The ultimate goal of the economy is defined as maximization of aggregate income or wealth.

As Hamlet says „There are more things in heaven and earth, Horatio, Than are dreamt of in your philosophy". (Hamlet 1.5.167–8). The economic reality in the twenty-first century is far more complex than it is depicted in mainstream economics today. Consequently a much richer and more sophisticated ontology is need in economics to increase its interpretative and predicting power.

11.2 Variety of Economic Actors

In today's economic life we can meet a variety of actors. A tentative list of them is as follows:

businesses (corporations, unincorporated businesses, social enterprises)
financial institutions (central banks, commercial banks, investment founds, pension funds, insurance companies)
public service organizations (education, health care, utilities)
households (super rich, upper class, middle class, poor, marginalized)
local communities (cities, towns, villages)
virtual communities (facebook, linkedin, twitter)
political bodies (parliaments, central and local governments, international organizations)
legal institutions (courts of justice, law firms)
cultural institutions (churches, universities, R&D organizations, art institutions)
media (print, online, and audio-visual)
civil society organizations (environmental, human rights, cultural)
professional associations (chambers of commerce, etc.)
trade unions
clusters
robots (financial, physical)

These actors can be grouped into broad categories like individuals, organizations, communities, networks, and robots. They differ ontologically, which means that their motivations, cognitive capacities and moral capabilities are different.

Mainstream economics uncritically presupposes that each and every economic actor follows the same pattern of behavior, namely *maximize its self-interest*. This claim is empirically false. Behavioral economists collected overwhelming empirical evidence which convincingly demonstrates that economic actors systematically violate the requirements of the Homo Oeconomicus model. They display bounded rationality, bounded willpower and bounded self-interest. Bounded rationality reflects the limited cognitive abilities that constrain human problem solving. Bounded willpower captures the fact that people sometimes make choices that are not in their long-run interest. Bounded self-interest incorporates the comforting fact that humans are often willing to sacrifice their own interests to help others. (Mullainathan and Thaler 2000; Thaler 2015)

Economic behavior is multifaceted and context-dependent. (Zsolnai 2013) Richer behavioral models are needed which can capture the self-regarding, other-regarding and norm-regarding aspects of choice behavior of economic actors. (Zsolnai 2008)

11.3 Coordination Mechanisms: Crowding In or Crowding Out?

Based on his life-time preoccupation with non-equilibrium economic system Janos Kornai (Kornai 1984, 1990, 1992) identified four basic coordination mechanisms which play decisive role in the working of economic systems. Coordination is defined by Kornai as the regulation of two or more interacting economic actors. The four coordination mechanisms are the market coordination, the bureaucratic coordination, the ethical coordination, and the aggressive coordination.

Market Coordination There is a horizontal relationship between the buyer and the seller. They enter into the relationship voluntarily. Each of them is motivated to make a profit or surplus. The transaction is monetarized.

Bureaucratic Coordination There is a hierarchical relationship (sub- and superordination) between the actors. The participation in the relationship is non-voluntary. Subordinate actors are motivated to accept the orders and prohibitions of the superordinate actors through administrative coercion supported by legal sanctions. The relationship is lasting and institutionalized.

Ethical Coordination Horizontal relationship exists between the actors. They enter into the relationship on voluntary basis. Coordination may be based on reciprocity or one-sided altruism. The aim of the interaction is not material gain but help the others.

Aggressive Coordination There is a hierarchical relationship between the superordinated and the subordinated actors. Participation is non-voluntary. A willful force is used by the superordinated toward the subordinated to achieve the desired transaction.

Kornai (1984) notes that in reality the different coordination mechanisms operate side by side. Their scope is partly disjunct but they assert themselves partly intertwined. Combinations of the coordination mechanisms occur. However, the combination of different mechanisms is usually not stable. Bureaucratic or aggressive coordination may crowd out market or ethical coordinations. Also, market coordination may crowd out ethical coordination.

In the psychology literature, Deci (1976) and Deci and Ryan (2000) have illuminated the relationship between intrinsic and extrinsic motivation, i.e. motivation that is inherent in the task itself and motivation that relies on external rewards or sanctions. In the economic literature, the crowding-out effect, i.e. instances where external rewards in fact reduce the internal motivation of the actor, have been explored both conceptually and through a number of experimental studies (Frey 1997).

In his book "The Moral Economy" Samuel Bowles (2016) extensively documents and analyses the dynamic interaction between market forces and ethical motivations.

An instructive case is the famous experiment of the *Haifa's day care center* (Gnezzy and Rustichini 2000). In Haifa at day care centers experimenters introduced a fine for parents to reduce their late arrival for picking up their children. The result was disastrous. Parents responded to the fine by doubling the late arrival. After twelve weeks the experimenters revoked the fine but the parents' higher level tardiness persisted.

Analyzing the Haifa case Samuel Bowles (2016, 4–5) concludes that the counterproductive result suggests "a negative synergy between economic incentives and moral behavior. Placing a price on lateness, (...) seems to have undermined the parents' sense of ethical obligation to avoid inconveniencing the teachers, leading them to think of lateness as just another commodity they could purchase."

Another important experiment about the negative synergy between market incentives and ethical motivation was executed by Armin Falk and Nora Szech (2013). They demonstrate that using the market as a frame of reference in decision making provides a good excuse for people to morally disengage from considerations of doing harm and damage for third parties. In the experiment executed by Falk and Szech subjects decided between either saving the life of a mouse or receiving money. In market setting, the willingness to kill the mouse was substantially higher than in the case of individual moral decisions.

There are cases where ethical motivation and market incentives go hand in hand and create positive synergy. Such a insightful case is the *Irish plastic bag tax* case.

In 2002 the government introduced a small tax on plastic bags. This market incentive was preceded by a huge publicity campaign to show the dramatic effect of plastic bags in the natural environment. The effect was phenomenal: in two weeks the use of the plastic bags in Ireland dropped by 94%. (Rosenthal 2008) "The tax may have crowded in social preferences: for many Irish man and women, carrying a plastic grocery bag home appeared to have joined wearing a fur coat in the closet of antisocial practices. (...) the monetary incentive was combined with a message of explicit social obligation, and it apparently reminded people of the larger social cost of using and disposing of the bag." (Bowles 2016, 202–203.)

The negative and positive synergies between the different coordination mechanisms suggest that the fabric of the market economy is a highly complex, adaptive system where incentives, constraints and preferences play a dynamic role: sometimes undermine, sometimes reinforce each other.

11.4 Substantive View of the Economy

Mainstream economics suggests that the main goal of economic activities is to produce material gains. This implies that the ultimate goal of the economy is to maximize aggregate income or wealth.

Aristotle made an important distinction between oikonomia and katalaxia. *Oikonomia* is about activities to satisfy material needs of the family while *katalaxia* is to produce goods, especially food, for trade outside the community. Oikonomia has a limiting principle since the material needs of the family are finite. Contrary to this, katalaxia has no limiting principle because it is driven by the commercial activity of moneymaking. While oikonomia concerns sufficiency and katalaxia concerns maximization of monetary gains.

Following Aristotle *Karl Polanyi* formulated a substantive theory of the economy. He distinguished between the formal and the substantive meanings of the term "economic". (Polanyi 1977)

The *formal* meaning of "economic" springs from the logical character of the means-ends relationship; as in economizing or economical; from this meaning springs the scarcity definition of economic. Contrary to this, the *substantive* meaning of "economic" points to the elemental fact that human beings, like all other living things, cannot exist for any length of time without a physical environment that sustain them; this is the origin of the substantive definition of economic.

The substantive meaning stems from man's patent dependence for his or her livelihood upon nature and his or her fellows. Man survives by virtue of an institutionalized interaction between himself and his natural surroundings. That process is the economy, which supplies man with the means of satisfying his or her material needs. So in substantive sense "economic" denotes nothing else than bearing reference to the process of satisfying materials needs of the community.

Polanyi (1977) suggests that throughout history the main goal of economic activities is always to provide livelihood for those who are involved. This implies that the ultimate goal of the economy is to create *sustainable* and *just prosperity* against which the performance of the economic system should be measured.

Prosperity is not identical with aggregate income or wealth measured by GDP or GNP. Today, countries around the world judge economic performance by GDP, which measures the market output of goods and services. However, GDP ignores all activities outside the marketplace, and so it provides a lopsided and incomplete evaluation of economic performance. (Brown 2016, 104)

More holistic measure of the performance of the economy have already been developed. Among them are Genuine Progress Indicator, Bhutan's Gross National Happiness Index, the OECD Better Life Index, and the Happy Planet Index.

Genuine Progress Indicator is a single-value measure which expands on GDP by adding in the value of nonmarket benefits (e.g., child rearing, leisure) and subtracting the value of nonmarket costs (e.g., environmental degradation, depletion of natural capital), and also adjusting for inequality across the population.

Bhutan's *Gross National Happiness* (GNH) Index us a non-monetary measure. It is a complex summation of nine equally important domains, namely psychological well-being, use of time, community vitality, cultural diversity, ecological resilience, standard of living, health, education, and good governance.

OECD Better Life Index (BLI) uses a dashboard of indicators: income, jobs, community, education, housing, environment, governance, health, life satisfaction, safety, and work-life balance. Based on surveys that ask people their perceptions about their lives, health, and government.

Happy Planet Index (HPI) measures a country's happiness score times life expectancy (both adjusted for inequality) divided by its ecological footprint (the amount of land required per person to replicate the country's consumption over time).

A simple but powerful measure of prosperity can be generated as follows:

P = (S, H)

where „S" refers to sustainability while „H" refers to human well-being.

This two-component measure does not presupposes any substitutability between sustainability and human well-being. Rather it reflects the possible trade-offs between these variables. Indeed there are serious trade-offs between sustainability and human well-being in today's economies.

There are relatively simple indices which can be used for determining the components of prosperity. Sustainability can be measured by the ecological footprint while well-being can be measured by the human development index.

Ecological footprint measures the ecological assets that a given population requires to produce the natural resources it consumes (including plant-based food and fiber products, livestock and fish products, timber and other forest products, space for urban infrastructure) and to absorb its waste, especially carbon emissions. Ecological footprint is expressed in global hectares—globally comparable, standardized hectares with world average productivity. (Global Footprint Network 2017)

The well-known *Human Development Index* (HDI) is a composite index of life expectancy, education, and per capita income. A country scores higher HDI when the lifespan is higher, the education level is higher, and the GDP per capita is higher. (UNDP 2010) A more sophisticated version of HDI is the so-called Inequality-adjusted Human Development Index (IHDI) which also account for inequality. IHDI measures of the average level of human development of people in a society once inequality is taken into account. (UNDP 2010)

Data collected by the Global Footprint Network suggest that very few countries are achieving high human development (HDI 0.67 or higher) within a globally replicable level of biocapacity demand (per capita Footprints lower than 1.79 global hectares for 2007). Moran et al. (2008) observe that as countries improved human well-being, their resource use grew. Beyond a certain level, small HDI gains are likely only obtainable via large ecological footprint increases.

A recent study by the European Environmental Agency (2015) show that there is no country today which provides high level human well-being while staying within the world biocapacity demands.

To achieve the sustainable prosperity requires other actions and policies than those recommended by mainstream economics. Ecological economists suggest that prosperity can be attained without economic growth. Tim Jackson (2017) forcefully argues that the economy of tomorrow may be transformed in ways that protect employment, facilitate social investment, reduce inequality and deliver both ecological and financial stability.

11.5 Conclusion

The ontological assumptions of mainstream economics should be reconsidered. Better ontology can provide a basis for developing more ethical and effective economic actions and policies.

In the twenty-first century economy economic actors represent diverse types of beings and the monetized transactions are only one form of the interactions among them. Economics should minimally consider the following types of economic actors: individuals, organizations, communities, networks, and robots (or algorithms). Among the economic actors different types of interactions can be identified: market exchange, bureaucratic coordination, ethical transaction, and aggressive action. In reality different types of interactions are usually combined but "crowding in" and "crowding out" effects may happen among them. For example, market exchange may crowd out ethical coordination or bureaucratic coordination may crowd out market exchange.

Taking a more universal perspective the ultimate goal of the economic system is not to produce material gains but to provide livelihood for those who are involved. In this way sustainable and just prosperity can be considered as the ultimate goal of the economy against which the performance of any economic system should be measured.

References

Bowles, S. 2016. *The moral economy. Why good incentives are no substitute for good citizens*. New Haven /London: Yale University Press.

Brown, C. 2016. *Buddhist Economics. An Enlightened Approach to the Dismal Science*. New York/London/Oxford/New Delhi/Sydney: Bloomsbury Press.

Deci, E.L. 1976. The hidden cost of rewards. *Organizational Dynamics* 4 (3): 61–72.

Deci, E.L., and R.M. Ryan. 2000. The "what" and "why" of goal pursuits: Human Needs.and the Self-Determination of Behavior. *Psychological Inquiry* 11 (4): 227–268.

European Environmental Agency. 2015. *Setting the scene*. https://www.eea.europa.eu/soer-2015/global/setting-the-scene. Accessed 23 Oct 2017.

Falk, A., and N. Szech. 2013. Morals and markets. *Science* 340 (2013): 707–711.

Frey, B. 1997. *Not just for the money: An economic theory of personal motivation*. Cheltenham: Edward Elgar.

Global Footprint Network. 2017. *Ecological footprint*. https://www.footprintnetwork.org/our-work/ecological-footprint/. Accessed 23 October 2017.

Gnezzy, U., and A. Rustichini. 2000. Pay enough or Don't pay at all. *Quarterly Journal of Economics* 115 (2): 791–810.

Jackson, T. 2017. *Prosperity without growth. Foundations for the economy of tomorrow*. 2nd ed. London/New York: Routledge.

Kornai, J. 1984. Bureaucratic and market coordination. *Osteuropa Wirtschaft* 29 (4): 306–319.

———. 1990. The affinity between ownership forms and coordination mechanisms. The Common Experience of Reform in Socialist Countries. *Journal of Economic Perspectives*., Summer 1990 4 (3): 131–147.

———. 1992. *The Socialist Sytem*. Princeton/Oxford: Princeton University Press /Oxford University Press.

Mann, R. 2010. *An introduction to particle physics and the standard model*. Boca Raton: CRC Press.

Moran, D., M. Wackernagel, J. Kitzes, S. Goldfinger, and A. Boutaud. 2008. Measuring sustainable development — Nation by nation. *Ecological Economics* 64: 470–474.

Mullainathan, S. and Thaler, R.H. 2000. *Behavioral Economics*. The National Bureau of Economic Research. Working Paper No. 7948. http://www.nber.org/papers/w7948. Accessed 23 Oct 2017.

Polanyi, K. 1977. *The livelihood of man*. New York: Academic.

Rosenthal, E. 2008. Motivated by a tax, Irish spurn plastic bag. *New York Times* February 2, 2008.

Thaler, R.H. 2015. *Misbehaving. The making of behavioral economics*. New York/London: W.W. Norton.

UNDP. 2010. *The human development concept*. New York: United Nations Development Program.

Zsolnai, L. 2008. *Responsible decision making*. New Brunswick/London: Transactions Publishers.

———. 2013. The moral economic man. In *Handbook of business ethics – Ethics in the new economy*, ed. Laszlo Zsolnai, 35–53. Oxford: Peter Lang International Academic Publishers.

Part V
Implications for Economic Policy

Chapter 12
How Does the Methodology of Neoclassical Economics Eliminate the Question of Fairness?

Zoltán Pogátsa

Abstract Neoclassical economics is almost unique amongst the different schools of economics (mostly hidden not only from public debates, but also from graduate education) in that it does not address the question of distribution. By ignoring this important aspect, it avoids one of the important ethical questions of fairness and social justice. Mainstream ideology of capitalist societies, neoclassical economics, assumes a fair distribution and flow of resources across society, and ostracizes and intimidates those who are critical about this assumption. The most puzzling aspect of neoclassical theory is that in perfectly competitive markets there can be no profit. As we all know, competition eliminates the profit margin by reducing the unit sales price to the level of unit costs. Thus a key epistemological axiom of neoclassical economics, perfectly competitive markets, is at odds with another axiom, namely that profits drive investment and entrepreneurship.

Neoclassical economics became mainstream in the twentieth century in terms of academic research, university teaching, as well as the public perception of economics, mainly through the media. However, neoclassical economics is almost unique amongst the different schools of economics (mostly hidden not only from public debates, but also from graduate education) in that it does not address the question of distribution. By ignoring this important aspect, it avoids one of the most important ethical questions, that of fairness and social justice. (The other being sustainability.)

Starting from the almost universally held ideal that human beings are equal, it is possible to construct a normative framework for an actually functioning economic system. It should be one where every human being, regardless of class and identity based background should have equal chances of competing, and their success should depend only on their individual effort. Not even on their performance, since that also depends on preconditions beyond our control, for instance personal strength.

Z. Pogátsa (✉)
University of West Hungary, Sopron, Hungary
e-mail: pogatsa.zoltan@uni-sopron.hu

© Springer International Publishing AG, part of Springer Nature 2018
P. Róna, L. Zsolnai (eds.), *Economic Objects and the Objects of Economics*, Virtues and Economics 3, https://doi.org/10.1007/978-3-319-94529-3_12

We acknowledge that it would be possible and perhaps even desirable to imagine an economic system that is not meritocratic, but is built for instance on needs, but here we would like to demonstrate that even within the context of a capitalist economy it is possible to discuss fairness.

Personal success should not depend on factors that are inherited and are beyond the control and therefore the sphere of responsibility of the individual, such as wealth. In fact John Rawls comes close to formulating such normative criteria with this *Theory of Justice* (Rawls 1971), although he concentrates on redistribution, and does not address fairness in the process of the creation of value. (The main line of criticism on Rawls, Nozick's attack (Nozick 1974), is feeble. It presupposes that rewards in the economy are entirely a return on one's individual effort, to which society contributes nothing. This is clearly untenable. One's ability to provide supply is clearly dependent on socially provided prerequisites, such as education, public order, health, transport infrastructure, etc.)

Thus in spite of the relativistic era ushered in by postmodernism, the concept of justice formulated in the above way enjoys very wide support. Therefore it is relevant to apply this normative framework to the existing economic system. Indeed authors such as Thomas Piketty (2014), Emmanuel Saez (see Piketty et al. 2018), Branko Milanovic (2011) and others have done so successfully recently, bringing the question of inequality, justice and (re)distribution back into focus of economics, concluding that without strong trade unions and a high degree of redistribution, "market coordination" is not a fair and just system. It does not function as a level playing field "catallaxy", as Hayek would have us imagine it (von Hayek 1978: 108). It is more a latent process of concentration of resources.

Why did Piketty's (2014) book come as such a revolutionary input? How can it be, that economics as a discipline failed to monitor the concentration of wealth, incomes, capital and other resources for so long? How could economics, as well as its projection into public discourse, rest comfortably on the presumption that the economy is a neutral playing field, where we do not need to examine the initial distribution of capital in order to assume fair competition? How could economics develop into the most dominant social science without even applying any criteria of justice of fairness?

The main explanation to these troubling questions is to be found in the underlying epistemology of neoclassical economics. This dominant school of the discipline eliminates questions related to the distribution of resources through the very construction of the abstract model it erects.

For starters, this model relies on methodological individualism, which disregards the fact that individuals might and almost always do rely on resources available to a larger community. The best but certainly not the only example of this is the family.

As a next step, neoclassical economics simplifies the very complex preference structures of society into a single representative agent that signifies the costumer. In order to be able to make the (epistemologically illegitimate) move from microeconomics to macroeconomics,[1] it needs to reduces the extreme complexity of the real economy to a theoretical abstraction where its entirety is represented a single consumer. Furthermore, there is also a single representative factory on the supply side. One of the perhaps unintended, but definitely "useful" consequences of this move of reducing a complex society into a single individual is to ignore issues related to assumptions about the starting distribution of resources across society, as well as all consequent processes of the flow of resources thereafter. In the neoclassical model there are no social classes that compete for resources, not even the social layers of a non-Marxian approach, a less relational term.

Steve Keen's summarizes the absurd essence of the neoclassical model as follows:

> ...a model of the macroeconomy as consisting of a single consumer, who lives for ever, consuming the output of the economy, which is a single good produced in a single firm, which he owns and in which he is the only employee, which pays him both profits equivalent to the marginal product of capital and a wage equivalent to the marginal product of labor, to which he decides how much labor to supply by solving a utility function that maximizes his utility over an infinite time horizon, which he rationally expects and therefore correctly predicts. The economy would always be in hey didn't see it coming equilibrium except for the impact of unexpected 'technology shocks' that change the firm's productive capabilities (or his consumption preferences) and thus temporarily cause the single capitalist/worker/consumer to alter his working hours. Any reduction in working hours is a voluntary act, so the representative agent is never involuntarily unemployed, he's just taking more leisure. And there are no banks, no debt, and indeed no money in this model. (Keen 2011: 257–8)

It would be very useful for general public discourse to have all citizens receive some training in economics, in order for them to be reflective and aware of the ridiculous assumptions of the dominant neoclassical model on which the "vulgar economics" of the media is built. It is unlikely, however, that the general public will ever access such a level of awareness about the epistemology of the very essence of the worldview that effectively directs their lives. It remains passively useful for social peace not to problematise the assumptions that underpin this defining ideology of our age, as effectively every other school of economics that would replace the neoclassical one would include distributive issues and therefore questions about social justice. Neoclassical economics is very convenient, as it professes to be value free. We clearly have a case of the elimination of the ethical questions in economics through (flawed) epistemology.

A further significant characteristic of neoclassical economics is that it concentrates on the sales markets of already existing goods. As symbolized by its central epistemological element, Marshallian supply and demand curves (Marshall 1890), neoclassical economics focuses on markets for goods that do not need to be produced, but are somehow given, readily available. By doing so, it eliminates a further important issue, the creation of value. This is a crucial move again, since economics

[1] Illegitimate, because it ignores the principle of emergent properties. Micro phenomena cannot be transferred to macro phenomena without taking into account the interactions between them, which may result in completely new characteristics emerging at the macro level.

has been struggling with the question of value ever since its early stages. It is debatable whether consistent theories of value have indeed since been established (Varoufakis et al. 2011), but to ignore the challenge altogether is to avoid once again the moral normativity of social justice and fairness. Who produces value? How? Are they fairly rewarded for it? These are question that unavoidably arise when economics does not turn its behind on the world of production rather then commerce.

"Merchant profit" is easily explainable as a monopoly profit arising from the opportunity for arbitrage. But how about the non-monopoly profit of perfectly competitive markets?

The most puzzling aspect of neoclassical theory is that in perfectly competitive markets there can be no profit. As we all know, competition eliminates the profit margin by reducing the unit sales price to the level of unit costs. Thus a key epistemological axiom of neoclassical economics, perfectly competitive markets, is at odds with another axiom, namely that profits drive investment and entrepreneurship.

Should we then assume that all profit is monopoly profit? Various theories have been advanced to avoid this problematic conclusion. Frank Knight's famous formulation (Knight 1933) as profit accruing to capital accepting unknown risks leads to nothing other than hazard and chancing. If profits are not rewards for the conscious efforts of capitalists, then the economic system is reduced to a casino, and entrepreneurs to gamblers. Schumpeter's formulation (Schumpeter 1942) about innovative monopoly position is acceptable, but poses two separate challenges. Firstly, it does not explain the profits of non-innovative enterprises. Secondly, it does not explain why profits arising from innovation are not accrued to engineers, inventors, designers, who are paid a salary, but to shareholders, who were had not been involved in the innovation in the first place.

Austrian theories of profit imagine profit as arbitrage. This is clearly unreconciable with the theory of perfectly competitive markets. Kirzner's formulation (Kirzner 1979) of production being equal to arbitrage of components is plainly wrong. The market of the final product is not the same market as the market of the components, and arbitrage presupposes homogenous products. Production involves the labor process, whereas arbitrage (merchant profit) does not.

Non-neoclassical theories have alternative resolutions of this problem. Marx (1954) famously finds profit in perfectly competitive situations, arising out of exploitations. In other words in his system the market for goods might be allowed to be perfectly competitive for a ceteris paribus hypothetical situation, but the market for labor is not. He famously runs into the transformation problem, and it is debatable whether this challenge has been resolved.

Neo-Ricardian economists (Sraffa 1960) demonstrated in the Cambridge Capital Controversies (Harcourt 1972) that the neoclassical notion of profit and wages being the reward for the marginal productivity of capital and labor respectively is false. What sort of moral challenge arises from this finding? And why is it that the mainstream of the economics profession, consisting of neoliberal economists and pundits, is allowed to continue to ignore the falsification of marginal productivity theory? Does it not serve the class purpose of maintaining the morally relativist social status quo, in order to avoid making the non economically trained majority aware that distributing value-added according to falsely hypothesized marginal productivity theory is epistemologically and therefore ethically wrong?

Soviet style communism was famous for suppressing enquiries about social structure. The official position of the system was that there is absolute equality in the workers state, and those questioning this declared state of affairs found themselves ostracized or intimidated. Similarly, the mainstream ideology of capitalist societies, neoclassical economics, assumes a fair distribution and flow of resources across society, and ostracizes and intimidates those who are critical about this assumption.

References

Harcourt, G.C. 1972. *Some Cambridge controversies in the theory of capital*. Cambridge: Cambridge University Press.

von Hayek, F.A. 1978. *Law, legislation and liberty*. Chicago: University of Chicago Press.

Keen, S. 2011. *Debunking Economics: The Naked Emperor Dethroned*. London/New York: Zed Books.

Kirzner, I. 1979. *Perception, opportunity and profit: Studies in the theory of Enterpreneurship*. Chicago: University of Chicago Press.

Knight, F.H. 1933. *Risk, uncertainty, profit*. London: London School of Economics and Political Science.

Marshall, A. 1890. *Principles of economics*. London: Macmillan.

Marx, K. 1954. *Capital Vol. 1*. Moscow: Foreign Languages Publishing House.

Milanovic, B. 2011. *The haves and the have Nots: A brief and idiosyncratic history of global inequality*. New York: Basic Books.

Nozick, R. 1974. *Anarchy, state and utopia*. New York: Basic Books.

Piketty, T. 2014. *Capital in the Twenty-first Century*. Cambridge, Massachusetts/London: The Belknap Press of Harvard University Press.

Piketty, T., E. Saez, F. Alvaredo, and L. Chancel. 2018. *The world inequality report*. Cambridge, Massachusetts: Belknap Press.

Rawls, J. 1971. *A Theory of Justice*. Cambridge, Massachusets: The Belknap Press of Harvard University Press.

Schumpeter, J. 1942. *Capitalism, socialism and democracy*. New York: Harper and Brothers.

Sraffa, P. 1960. *The production of commodities by means of commodities*. Cambridge: Cambridge University Press.

Varoufakis, Y., J. Halevi, and N.J. Theocarakis. 2011. *Modern political economics: Making sense of the Post-2008 world*. London/New York: Routledge.

Chapter 13
Economics Is a Moral Science: A Value Based Approach

Arjo Klamer

> *I see us free, therefore, to return to some of the most sure and certain principles of religion and traditional virtue—that avarice is a vice, that the exaction of usury is a misdemeanor, and the love of money is detestable, that those walk most truly in the paths of virtue and sane wisdom who take least thought for the morrow. We shall once more value ends above means and prefer the good to the useful. We shall honor those who teach us how to pluck the hour and the day virtuously and well, the delightful people who are capable of taking direct enjoyment in things, the lilies of the field who toil not, neither do they spin.*
>
> *John Maynard Keynes. Essays in Persuasion, p 161*

Abstract Keynes prophesies economics as a moral science. Economists would abandon instrumentalist orientation where means get all our attention and the ends virtually none, and focus on what is the good life and the good society. The paper advances to that end the value based approach. It focuses on two interventions that seem to be necessary to realize economics as a moral science. One is to redefine economics and the second is to expand the economic understanding of goods. Defining economics as the discipline that studies the realization of values is the first step to perceive economics as a moral science. The next step is the inclusion of shared goods as an economic category. The result is a broader scope of economics, one that allows us to consider the goods and values that are of ultimate importance to people, groups, organizations and societies. The notion of shared good, combined with the notion "willingness to contribute" furthermore addresses the problem that standard economics has with altruistic behavior. The paper argues that most of our actions are not directed at private gain, or the acquisition of private goods, but at the realization of shared goods.

A. Klamer (✉)
Erasmus University Rotterdam, Rotterdam, Netherlands
e-mail: klamer@eshcc.eur.nl

© Springer International Publishing AG, part of Springer Nature 2018

P. Róna, L. Zsolnai (eds.), *Economic Objects and the Objects of Economics*, Virtues and Economics 3, https://doi.org/10.1007/978-3-319-94529-3_13

In this famous statement Keynes prophesies economics as a moral science. We economists would cease to be cynical about the intentions of individual agents and focus on doing the right thing. We would abandon our instrumentalist orientation where means get all our attention and the ends virtually none, and focus on what makes people truly happy. In other words, economics would be the discipline concerned with the good life and the good society, just as it was before the economics of Lionel Robbins, Jan Tinbergen and Paul Samuelson.

In my book "Doing the Right Thing" (Klamer 2017) I have tried to figure out what it takes to transform the discipline into a moral science. I advance to that end the value based approach. Here I will focus on two interventions that I deem necessary to realize economics as a moral science. They constitute important elements of the value based approach. One is to redefine economics and the second is to expand the economic understanding of goods. But let me first briefly motivate the need to take Keynes' prophesy seriously.

13.1 Why Should We Conceive Economics as a Moral Science?

The most important reason is that economics already functions as a moral science. Another reason is that everyday life consists to a greater part of moral actions; people, organizations and societies are trying to do the right things all the time. Economists only seem blind to the normative consequences of their practices and the value laden character of everyday life.

The blindness comes about because of the instrumentalist framing of the discipline. John Neville Keynes, Lionel Robbins and Jan Tinbergen, among others, made economists believe that they practice a 'positive' science focusing on "what is, preoccupied with the means leaving the thinking about the ends, about what" should be", to others. As Lionel Robbins puts it in his canonical 1932 essay:

> Economics [..] is concerned with that aspect of economic behavior which arises from the scarcity of means to achieve given ends. It follows that Economics is entirely neutral between ends [..]. Economics is not concerned with ends as such. (Robbins, 1932: 23)

Yet, the instrumentalist focus is not what makes economics politically effective. It is hard to detect the type of influence that Tinbergen (1956) had anticipated when he introduced his engineering perspective on the discipline of economics. Tinbergen's idea was that economic science would provide the necessary information that politicians need to make rational policy decisions. Such information was supposed to be scientific and hence objective. That is not how it works. It is actually difficult and mostly impossible to detect the engineering by economists. It is not that economists feed politicians with the results of their research and that politicians act upon those. In reality political decision making is complex, may involve outcomes of economic research, but more often exposes disagreements among economists and incorporates other arguments that are non-economical, such as particular interests and over-arching values. Economists often report feeling sidelined in important policy decisions (see for example Klamer 2003).

However, economics does not make its influence felt by way of its positive results; its main impact is by way of its normative content. This influence is especially felt in the current neo-liberal environment. Following Foucault and Sennelart (2008), I define neo-liberal the practice of governors and bureaucrats of public as well as private organizations who apply the logic of the market. This happens when government workers speak of welfare recipients as "customers," when bureaucrats stress the importance of "choice," "competition," "results," "cultural entrepreneurship," and "market pricing" of government services. When bureaucrats say and write these words, they do not refer to some kind of scientific study; they do so because they find all these elements important for policy. They constitute values for them. Accordingly, these policy makers adopt from elementary economics not the positive outcomes but the values that are implicit in an standard economic perspective.

Normative is also the impact of economics on life in general. The economic values show when people advocate and embrace "competition", stress the importance of "free choice," celebrate "entrepreneurship," trust "market processes," and recognize "opportunity costs" when a meeting lasts too long or when local people advocate turning an empty space into a playground.

In all these instances the adoption is not neutral, nor objective. When people value "competition" in the Dutch context they devalue the type of collaborations that have characterized the Dutch economy since people had to bind together to hold back the water. The stance is normative. When someone spends time with his family and brings up the notion of opportunity costs ("hey guys, do you know what I could have made in the time I spent with you"), it probably affects the appreciation of that time. The same would be the case when friends begin to charge each other for self-cooked meals "because that is more efficient." And what to say when people apply their economic lessons in an argument about the pros and cons of a market for kidneys and other human organs? The market logic of standard economics is in all these cases value-laden.

Everyday life is value laden, too. Take the dinner that you and your colleagues decide to have at a nice restaurant downtown. In our standard accounts we will focus on what we pay at the end; we call that "consumption," and are keen to know the income that gets generated. Those numbers appear in the national accounts. Yet is that what our dinner is about? I expect not. More likely you will have dinner because you value "collegiality", and the "conversation" that you will be having. If I were to ask you about your dinner, you most probably will appreciate the dinner for the quality of the conversation. If we define moral action all what we do to realize what is truly important for us, than the decision to seek a good conversation with colleagues could be considered a moral decision. It certainly is for academics, so I hope.

Even within the world of economists values seem to direct scientific behavior as where economists insists on mathematical modeling or where other economists stress the importance of an institutional approach. As I have argued throughout my scholarly life, major disagreements among economists are because of divergence of scientific values. Whereas some value scientific rigor and a formal approach, others find relevance more important.

In order to make sense of such an impact, it works better to acknowledge that economics is a moral science after all. And to do justice to the moral impact, we do better with a value based approach. At least, that is what I consider important.

13.2 Changing the Definition

In his essay Lionel Robbins (1932) takes on economists who define their discipline as the study of the causes of material welfare. Robbins sees the limitation in this definition when he considers an opera singer who does not contribute to material welfare. According to Robbins his singing is paid for and therefore part of the economy. He then relates the story of Robinson Crusoe to make the point that our time is limited and hence we need to choose between alternative things to do. Crusoe could, as Robbins suggests, talk to a parrot, practice philosophy, or tend to his garden. That brings Robinson to his more encompassing definition of economics as the discipline that "studies human behavior as a relationship between ends and scarce means which have alternative uses" (Robbins 1932: 15).

However, reading the story of Robinson Crusoe could lead to a different conclusion. His actions just as well could be interpreted as directed at the realization of things important for him. So after he figured out how to survive, he sought and found company in the person of Friday whom he rescued from being the dish in a cannibal's meal, and, more importantly, he searched for a reconciliation with his earthly father and his heavenly Father. (Crusoe is actually a Bildungs novel.) If we read the story that way, our perspective and with that our questions change rather dramatically. For one, the moral dimension of the story would emerge.

The leading question becomes: what is important to Crusoe? What is he pursuing after he finds himself all alone on an island? The answers surely help us understand his behavior, more so if we were just paying attention to his limited means. By speaking in terms of preferences, as economists who follow Robbins are used to do, the implicit question is about what people want. Asking what is important does more justice to the story of Crusoe. I suggest that it also makes more sense of basically everything we do. Acting in a context, or contributing to a context, be it that of your family, the organization in which you work, your profession, your city, your society or whatever entity, implies the realization of values, that is, of whatever is important. Whether I plan the family Christmas dinner, go to a theatre, shop in a market place, consider merging my organization with another, plan an exhibition, make art, deliver a speech, vote on a new law, or run with a friend, I am realizing values. Call such values ends, or purposes, they are what my actions are about. So why not taking them into account, or, to be consequential, to conclude that an economic study has to make explicit what values are at stake and how they are being realized, or not.

When I consider how to spend an evening, I need to take into account what is important to me. I could waste my time watching mindless television, but I also could consider spending time with my children or attending a theatre performance because the relationship with my children is important to me and so is being involved with theatre. When a theatre group decides to do something entirely different next year, it apparently seeks to realize something important to them. They may want to contribute something to the existing repertoire or bring about an innovation. They may also want to attract new visitors and generate additional income but those objectives are subsidiary to what really matters to them and that is to be an artistic group.

Is it possible that in everything we do, consciously and with the intent to feel good about it, we are seeking to realize something that is important to us? In all examples values are at stake, people are becoming or made aware of them, and then make them real by acting upon them. They may even reflect at a later stage to find out to what extent they succeeded. (When I lie down that evening I may conclude that, indeed, I wasted my evening with mindless television or had a great time with my children or at the theatre, or not.)

My suggestion is to interpret all such considerations and deliberations as the realizing of values in the sense of being aware of them, and the actions that follow as the realization of those values in the sense of valorizing them or making them real.

Realizing values has, therefore, two meanings: (1) being aware of one's values and those of others and (2) making them real, or valorizing them. It is the second part that is characteristic for the economic approach to the world. Where sociologists may be content with registering the values that people hold, economics, as I see the discipline, is also about their valorization. Valuing the relationship with my children is one thing, doing things to make that relationship valuable is quite another thing. Economics is about the activity that is valorization. Consequently, I suggest the following definition of economics: *Economics is the discipline that studies the realization of values, that is, it contributes to the awareness of the values that people, groups, organizations and societies hold and studies the ways in which values get valorized.*

Interpreted in such a way, economics is clearly positioned as a moral science, just as it actually functions.

When I use the term *valorization*, I refer to the making real of the relevant values. It is often interpreted as implying only the realization of exchange value (i.e. by selling something for a price), but I explicitly include the important values, such as artistic and social values. When artists produce a work of art, they will first have to realize its artistic values by bringing it in the conversation that is art. That means that they have to expose the work somehow, getting critics to view it and critique it. They will try to get the work in catalogues, on websites and, ideally, in important collections. Bringing the work to a gallery or an auction is only part of the valorization and in the case of art, not the most important part. Sure, by selling the work artists receive an amount of money that enables them to live and do things, but most important for the serious artists is to realize themselves as artist. And for that they need to valorize their work in the artistic world.

Focusing on the realization of values is what the cultural sector is all about, at least that is what, for example, the "serious" leaders of cultural organizations and "serious" artists are aspiring to do. An arts museum is dedicated to the arts, a theatre group to theatre. The artistic director of the Rotterdam Symphony orchestra wants to have the best musicians and the best conductor to perform Mahler's fifth as it never has been performed before. Or he wants to perform experimental music of a new composer because he truly beliefs in its power. His goal is to make great music.

In the pursuit of his goals this director of the orchestra may have to battle a lack of interest among regular visitors. He may have to deal with a skeptical business manager, and reluctant musicians. And he has to face the overwhelming preference for the usual, for recognizable music that makes listening easy. That is, he is up against the desire for amusement and entertainment. And because entertainment sells tickets, he has to be creative in realizing his music while avoiding bankruptcy.

With this value based approach to the subject of economics I follow a rich tradition to which a great range of authors and endless practices have contributed. I name the works of Thomas Aquino and Adam Smith, and, more recently, the works of Deirdre McCloskey. A favorite source of mine, though, is Aristotle and then in particular his Nicomachean Ethics.

13.3 Acting Upon Values Implies the Striving for the Good

Aristotle (384–322 BC) is the pragmatic Greek philosopher who would have a profound influence on civilization. That influence began in the 12th and thirteenth century when scholars translated his work into Latin. Aristotle's works became a major source for philosophers till the twentieth century. Thomas Aquinas (1225–1274) used the bible and Aristotle, to whom he referred as the Philosopher, as his main references, when he addressed moral issues in economic situations. Aristotle was more or less ignored during the major part of the twentieth century but is now back in vogue. Alasdair MacIntyre and Martha Nussbaum are two prominent contemporary philosophers who have brought his work back to life. In particular Nicomachean Ethics receives a great deal of attention nowadays, after having been ignored for a century or so. There are now even business handbooks that instruct managers how to apply Aristotle in their work.

It may be interesting to realize that Aristotle probably wrote the text to instruct his son Nicomachus. He began the instruction as follows:

> Every art and every inquiry, and similarly every action and pursuit, is thought to aim at some good; and for this reason the good has rightly been declared to be that at which all things aim. (Aristotle 350 BC, p. 1094a)

A contemporary reader may experience some difficulties here. Especially the notion of the good triggers questions and doubts–as if there would be a good out there for you and me to be realized. I do not want to pursue the philosophical discussion at this point and suggest staying pragmatic by interpreting the good as the purpose that a person, a community or an organization is seeking.

In an organization the purpose is expressed in the mission that some organizations have articulated and most have not.

Aristotle continues by suggesting that the good varies.

> But a certain difference is found among ends; some are activities, others are products apart from the activities that produce them. Where there are ends apart from the actions, it is the nature of the products to be better than the activities. (Idem, 1094a)

Notice that a good can be an activity. An actor wants to be a good actor. His good is in the acting, in rehearsing and in performing on stage; by acting he can realize his good. A craftswoman whose craft is making hats seeks to produce great hats. Her craft is a good and an activity at that.

Aristotle affirms such a reading when he continues.

> Now, as there are many actions, arts, and sciences, their ends also are many; the end of the medical art is health, that of shipbuilding a vessel, that of strategy victory, that of economics wealth. But where such arts fall under a single capacity- as bridle-making and the other arts concerned with the equipment of horses fall under the art of riding, and this and every military action under strategy, in the same way other arts fall under yet others- in all of these the ends of the master arts are to be preferred to all the subordinate ends; for it is for the sake of the former that the latter are pursued. It makes no difference whether the activities themselves are the ends of the actions, or something else apart from the activities, as in the case of the sciences just mentioned. (Idem 1095a).

Aristotle admonishes his son to distinguish means from ends. At the same time he urges him to be aware of the ends. It is a lesson that seems pertinent in a wide variety of situations today. It is a lesson for business managers who mistake the instrument of profit as an end in and of itself, or youngsters who seek money (lots of it!) as their goal. Swapping instruments for goals is all too common. A director of an American art museum declared in a seminar for art managers that he has three goals and they are (1) fundraising, (2) fundraising, and (3) fundraising. That sounds tough and he probably tried to unnerve his audience. Even so, he might be asked in the spirit of Aristotle what purpose the fundraising serves. It would be strange to set up a museum in order to raise funds. Such a goal also seems to make a bad proposition to those providing the funds. This director was swapping means for ends.

Asking others and yourself about goals is probing for the good that you and others are pursuing. It is the Aristotelian question. Money is never a goal or a good in itself. The pursuit of money is always a means to some end or another, even if it is sometimes difficult to articulate that end.

When we are aware of the goals, we need to figure out how to make them real or how to valorize them. To allow for a broader reach of the discipline we need a broader notion of what the goods are.

13.4 The Valorization of Values Involves the Realization of Goods, and Especially of Shared Goods

Say someone is desperate for love. That solves the awareness part; love is clearly an important value for this person. Yet seeking love is different from having it. In order to have love, this person has to do something and the least to do is to bring about a relationship with someone else. Without a relationship this person does not stand a chance of having love. It furthermore has to be a very special relationship in order to generate the love that this person is seeking. Here is the point: the relationship is a good in the sense that it is something to gain and something to loose. This person has a loving relationship, or not. It is a good as it is good for all kinds of values, like love. Accordingly, in order to make the value "love" real this person needs to acquire the good "a loving relationship".

Values need to be acted upon to become real, and an important way of realizing them is to acquire or generate goods. Goods enable us to realize values. They can be things like a cup of coffee or a computer, but they can also be intangible things, like relationships, communities, ideas, and artistic expressions. Goods are good for all kinds of things. You and I can "have" or possess goods, although often not in a legal sense. "Having" a relationship is the same as "possessing" it. People have houses and they have kids. Goods are whatever we have that enable us to realize values.

This characterization of goods may seem odd at first. It certainly is when you are used to think of goods as things you can hold onto, like a cup of coffee and a computer, or when you think of goods as things that you can buy or sell.

Again, Defoe's story of Robinson Crusoe helps to illustrate the point. When he was washed up on the shore of his island he took stock of his "assets" and "liabilities". This was his list:

Good	*Evil*
Still alive	Cast away
Singled out to be saved from the wreckage	Singled out to be miserable
I am not starved	Divided from mankind-solitaire
There are no wild beasts	I have no weapons
It is warm	I have no clothes

He could have added that he had been able to secure plenty of supplies from the shipwreck, including a few bibles (which serve him well in his search for his faith). He also found some coins in the ship but they are useless in his economy. (Note that economists latched onto a story that has no use for exchange! Strange, eh?)

His account stresses his social situation, or better the lack thereof. All alone he is barren from any social interaction. That is why his encounter with another human being after months of loneliness is so important to him. Friday is not only someone who can do his share of the chores, but he is also company for Crusoe. The latter teaches Friday some English so that they can *have conversations*. The conversations have a great deal of value for Crusoe, a social value. He feels better, and even richer because of them. Of course he does. We humans are in need of company and conversation. They are basic needs.

Were you to wear the glasses of demand and supply, exchange and price, costs and benefits, you do not see such things. With the value tinted spectacles the values of company and the conversations becoming pronounced. In a valued-based approach to economics friendship is a good. And so is a conversation. They resist having them. That means they do not come free; Crusoe has to make efforts and some sacrifices to acquire them. And they are good for all kinds of things.

13.5 What Are Goods?

In his Politics (1253b, 23–25) Aristotle defined goods as "the means of life and well being of men." So goods would be means to realize values, I would conclude. We are back to the definition I gave earlier. Carl Menger defined goods as something that has the capacity to satisfy identified human needs (Menger 2009, 1891). By stressing values I add an Aristotelian dimension; needs are less particular. A homeless man may say he needs booze but when we ask him what is important to him, he will probably say something like "belonging somewhere" or "having a family."

We also have to make clear that goods need to be acquired or generated and the acquisition requires some kind of sacrifice. To speak of "having" or "possessing" a good only makes sense if a good resists possession. When we happen to see a splendid sunset it would be strange to claim that sunset "ours". It would be different if you climbed a steep mountain in order to see the sunset: in that case you might say that it is yours to cherish. In general goods require an effort to acquire; we have to give something up to "have" them, or to enjoy the benefits of goods. Even a sunset requires some work in order to be able to enjoy it.

Therefore I would define goods as follows: *Goods are tangible or intangible things that an individual, a group of people, or a gathering of people possesses; they are good for all kinds of things, and their possession requires some kind of effort or sacrifice. Goods are important to us, because they enable us to realize values.*

If I were to follow Crusoe's example, I would try to assess what my activities have contributed to the goods I care for, such as the good "family", the good "collegiality", the good "knowledge", the good "a good conversation", the good "academic community", or the good "friendship. All those goods are important for me. They are good for a variety of my values.

Notice that these goods do not include the stuff I buy. All the things I buy are instrumental and usually serve the realization of the goods that are really important to me. The gas that I buy gets me to the places I need to be. The food I buy feeds me so that I can do my things, or it serves to accompany a conversation with colleagues or other people. By paying the interest on the mortgage I am able to share the house with my family and realize a home. The book I buy gives me access to potentially important knowledge. In short, the things I buy are of subsidiary importance; they are merely instrumental. The same is the case with the greater part of public goods that the state provides, with the exception for goods like "peace," and "justice" that I care about for their own sake.

Goods like "friendship", "knowledge," "art," and "good conversations" are unlike the goods that economists account for. They are most important to us, yet they are not included in the standard reasoning of economics. In order to allow for their inclusion we need an another category of goods. I call them "shared goods," because their ownership is shared and people have to contribute to them in order to claim (non-legal) ownership.

Let us investigate he good "a good conversation" to uncover the relevant characteristics of a shared good. It happens to be a good that is important to me: I am always looking for it, at home, with friends and at the university. It was also what Crusoe was looking for. There he was, all by himself, trying to survive, but with no one to talk to. In the end he was able to have conversations with this Friday guy. Even though they did not speak the same language, they quickly developed a language that they both could understand so that they could communicate. After some time they shared feelings, discussed what needed to be done and so on. Crusoe enjoyed the company and appreciated the ability of having a conversation with another person. The other presumably did as well.

The conversation that they are having is theirs; *they share the conversation.* It is a *joint production.* Friday makes an effort and so does Crusoe. A conversation in which he does not participate is not the same as one in which he is paying attention to what the other is saying, responds and contributes himself by telling things. The more effort they both put into their conversation, the better Crusoe will feel, and the better Friday most likely will feel as well. (He finds himself whistling again while walking along the beach and catches himself looking forward to the evening's meal because of the conversation they will be having.) It matters to him that Friday enjoys the conversation as well. They do not only produce the conversation together, they also "consume" it together. Its enjoyment is mutual. Their conversation is what he and Friday share. It is a shared good.

Maybe it is strange to call a conversation a good at all. From an economic point of view, however, it is not all that different from a good like bread. A conversation requires the input of time, effort and human and social capital. A conversation does not come about effortlessly. It is not free either. Even if the direct costs are zero, there will be opportunity costs (the benefits of alternative activities that are forsaken). And like bread a conversation gives satisfaction, a sense of companionship, knowledge, and, who knows, great insights. Thus far a shared good is like a private good. The conversation, however, cannot be a private good because no one can claim sole ownership of it. Crusoe cannot claim that the conversation is his and exclude others, including Friday, from enjoying it. A conversation is not *divisible.* We cannot say: "this part is mine and that is yours, I give you this for that." A conversation cannot be exchanged; it cannot be bought or sold. It is shared.

A conversation is not a public good either. A pure public good is both indivisible—when you and I are the public, I cannot consume it without your consuming it as well—and non-rivalrous—my consumption of the good cannot be at the expense of your consumption. Although a conversation cannot be split up and although Crusoe cannot have the conversation by excluding Friday, Friday and he can easily exclude others (like the members of Friday's tribe) from their conversation. Their conversation is between them; it is theirs. Maybe one of them will give it up for another conversation when the occasion presents itself. Crusoe may exclude Friday, as he actually did after being rescued. Conversations have rivalry and exclusion written all over them. A conversation, therefore, is a good that does accord neither with the class of private goods nor with that of public goods. It is a shared good. Both Friday and Crusoe enjoy having, or owning, it.

Note that this shared good is different from the notion of common goods in the standard economic discussions. Common goods are accessible to all but allow for rivalry. Economists tend to see great problems for the sustenance of common goods because of the free rider problem. Take the sea. This is a good that all people have in common. The whales swimming in the sea are a common good. Catching them is lucrative. Here the free rider problem occurs because whale hunters have an incentive to catch as many whales as they can, when they voluntarily agree to limit the number lest the whales die out, Individual whale hunters have the incentive to exceed that limit. They are said to be free riding (like what people do who do not buy tickets for public transport). The point is important since common goods lose their footing in the economic analysis because of this problem. But the problem does not apply to shared goods! When one party shirks, by pretending to be in the conversation while being with his thoughts somewhere else, the conversation will be different because of it and will have less value than a conversation to which all parties contribute. When there are more than two parties involved, one party may shirk and let the others do the work, but risk losing the conversation. When Friday and Crusoe are <having> a conversation, you may want to join in. Imagine they let you in and you subsequently shirk by not participating yourself. Apart from what they will think of your passivity, you will benefit differently from the conversation than they do, if you benefit at all. You may gain some information, some insight maybe, but you will not partake in the conversation and therefore it will not be <yours>. You can only have a conversation by partaking in it. (Of course, you can exploit a conversation that others are having, drawing gainful information from it, but then you cannot go home and tell your partner about this wonderful conversation you had.)

How to acquire a shared good? Take friendship as the example now. Like a good conversation, a friendship has shared ownership (of the friends involved), is costly to acquire (friends need to make sacrifices), and is good for the realization of all kinds of values. How do you acquire a friendship? It is not by way of a purchase: friendships are not for sale ("hey, I am too busy, want to take over a friend of mine?"). It is also not by some governmental program.

For a friendship it is important to share experiences, doing things together, and doing things for each other. That is what we agreed upon, too. "Friends have all things in common," the ancient Greeks would say. But that does not mean that having a friendship is a passive thing. Clearly, potential friends have to do something in order to acquire a friendship. And friends have to keep doing things to sustain or to strengthen a friendship. Aristotle pointed out that some degree of reciprocity is required. One friend does one thing for the other, and the other does something else at another time. They need to help each other, by giving a listening ear, helping the other to move, making dinner, giving support and so on. I will call all actions and gestures that serve the friendship somehow contributions to that friendship.

Contributions, therefore, are essential for the acquisition of a shared good. In an exchange situation people pay an amount of money in exchange of a private good. Economists speak of willingness to pay to indicate the willingness of people to pay a certain amount of money. Willingness to contribute indicates the willingness to contribute to a shared good. In this case there is no immediate return of something of equivalent values as in an exchange. When someone makes a contribution to a friendship, say by paying a sick friend a visit, the giving friend does not get anything in return except for the affirmation of the friendship. The receiving friend is made to understand that in order to sustain the friendship he will have to make a contribution somehow sometime later.

Willingness to contribute is quite different from willingness to pay. In the latter case the expectation is a return of equivalent value. In case of willingness to contribute the expectation is that the contribution will add values to a shared good.

13.6 Conclusion

Defining economics as the discipline that studies the realization of values is the first step to perceive economics as a moral science. The next step is the inclusion of shared goods as an economic category. One result is a broader scope of economics, one that allows us to consider the goods and values that are of ultimate importance to people, groups, organizations and societies. Often the ends that standard economics takes an unspecified given, turn out to be shared goods that require contributions of all kinds, outside the sphere of exchange, in order to get realized.

The notion of shared good, combined with the notion "willingness to contribute" furthermore addresses the problem that standard economics has with altruistic behavior. Thanks to these notions it becomes clear that most of our actions are not directed at private gain, or the acquisition of private goods, but at the realization of shared goods. When a soldier sacrifices his life, or when someone helps a homeless person they contribute to the good that we could name "a good society." Most so-called altruistic acts are contributions to one shared good or another.

Re-defining economics and the introduction of shared goods make for a good start of economics as a moral science and may turn out to do justice to the vision that Keynes so eloquently articulated.

References

Aristotle. (350 BC). n.d. *Nicomachean Ethics*. (W. Ross, Trans.) Retrieved from http://classics.mit.edu/Aristotle/nicomachaen.1.i.html

Foucault, M., and M. Sennelart. 2008. *The Birth of Biopolitics: Lectures at the College de France*, 1978–1979. Basingstroke: Palgrave MacMillan.

Klamer, A. 2003. *Speaking of economics: How to be in the conversation of economists*. London: Routledge.

———. 2017. *Doing the right thing: A value based economy*. London: Ubiquity Press.

Menger, C. 2009. *Principles of Economics*. Auburn: The Ludwig von Mises Institute Originally published in 1871.

Robbins, L. 1932. *An essay on the nature and significance of economic science*. London: MacMillan and Co.

Tinbergen, J. 1956. *Economic policy: Principles and design*. Amsterdam: North Holland.

Part VI
Conclusion

Chapter 14
Postscript on Ontology and Economics

Peter Róna

Abstract The proper domain of the discipline of economics then is the study of the ways and means with which individuals and groups decide what they should do to secure the material conditions of a worthwhile life. The requirement that the application of the theory must result in bringing about a worthwhile life rather than the maximisation or optimisation of any single issue also means that its tenets should be open to critical assessment on the basis of incommensurable considerations. Theories that lead to the degradation of human dignity or the environment will not do even if they satisfy the criteria of some algorithm. They are inadaquate not only because they are morally objectionable, but, more broadly, because they are ontologically deficient, that is to say, they fail to set out an adaquate picture of human reality.

There is something morally troubling about economics. Its objective to turn the study of how human beings secure the material conditions of their existence into a collection of law-like generalisations that are impervious to moral judgment constructs a regime under which hardly any of us would want to live even if it were possible. On top of it, the means, the methodology with which this reprehensible objective is pursued is never quite innocent of that certain whiff of intellectual sleight of hand. Unwarranted, empirically unjustified assumptions made in order to reach ideologically pre-selected conclusions is not the hallmark of intellectual integrity. What is one to make, for example, of the Arrow-Debreau general equilibrium model, constructed on the basis of seventeen assumptions fourteen of which cannot be true in the world as we know it? János Kornai (1971), for example, writes in his *Anti-Equilibrium* that this and similar Walrasian general equilibrium models purporting to provide an explanation of the price of commodities does nothing of the kind. Debreau' famous *Theory of Value,* as an example:

> … does not consider the relationship between its basic assumptions and axioms on the one hand and reality on the other; the question of whether the former reflect the latter exactly,

P. Róna (✉)
Blackfriars Hall, University of Oxford, Oxford, UK
e-mail: peter.rona@bfriars.ox.ac.uk

© Springer International Publishing AG, part of Springer Nature 2018

P. Róna, L. Zsolnai (eds.), *Economic Objects and the Objects of Economics*,
Virtues and Economics 3, https://doi.org/10.1007/978-3-319-94529-3_14

> approximately, or not at all is ignored. Nor is the question of how the deduced theorems relate to reality ever discussed. (Kornai 1971, 7)

Decorated with high mathematics as the means with which to command credibility, these and similar efforts do not even pause to specify the reality those numbers and formulas are said to represent, and do not explain how economic reality is turned into the sort of objects that can be mathematically represented. (Israel 1996) Indeed, an assumption is made, shared even by such illustrious critics of the general equilibrium theory as Kornai, that there *is* a reality that the techniques of modern economics can legitimately represent. However, the reality about which modern economics formulates generalisations on the basis of assumptions and axioms, is a contrivance, contrived to suit the assumtions and the models built upon those assumptions. Economic reality is not discovered because there is nothing to 'discover' in the sense in which the natural sciences make discoveries; it does not exist inderpendenly from human behaviour. It is *made,* contrived with the help of rules designed to guide, even coerce that behaviour. As further explained below, the proper subject of economics is how those rules and the roles constructed with the help of those roles are made.

It all went wrong with the Humean conceit, with the idea that facts can be separated from values in individual anf group social action. Modern economics is founded on this – fundamentally false – separation. As Isaiah Berlin (1996) put it:

> As any description of what is, embodies an attitude, that is, a view of it in terms of what should be: we are not contemplating a static garden; we are involved in a movement with a perceptible direction; it can be correctly or incorrectly described; but any description must embody a valuation, that is, a reference to the goals toward which the movement proceeds, and in terms of which it can be 'understood'… (Berlin 1996, 130)

Hume's dogma to the effect that 'no ought can be derived from an is' does not amount to much in the social sciences without a showing that there are 'ises' that do not contain an inseparable 'ought'. None of the logical positivists, such as Carnap and none of those who laboured to restore Humean empiricism, as, for example Ayer, troubled themselves with demonstrating that such 'ises' exist. (1) There are, of course, facts devoid of values, such the valences of the elements, but their factual purity is not the fruit of some operation that separates them from values, but, rather, the result of their independence from and imperviousness to human intentionality; they exist in a valueless existence, undisturbed by human consciousness and intentionality, unaffected by human judgment. But, as Berlin shows, life is an entirely different matter: "To live is to act. To act is to pursue goals, to choose, accept, reject, pursue, resist, escape, be for or against an entire form of life or some ingredient of it." (Berlin 1996, p. 130).

Humean fact are not the facts of life; they lack the prescriptive content inherent in objects of intentionality. It is also possible to generate synthetic facts, such as the rate of return on capital by contriving a notion of capital and the rate of return that excludes its qualitative aspects, but these 'facts' are neither the facts of the natural sciences, nor the facts of life. They are mimicking, but are unrelated to nature, and they are not encountered in ordinary life. They are sythesised surrogates contrived to occupy the space that facts would occupy in a hypothesis about nature. (2).

The evolution of modern social science – and most emphatically of economics – has been conditioned by its comparison with the natural sciences, by the apparent need to be successful in the terms that define and measure the success of the natural sciences. But the foundations for this comparison were procured at enormous cost, a long and tortuous process, in which the Humean separaction of 'fact' and 'value' was the essential first step.Upon the removal of value the process of reification with which events were deprived of their meaning could commence, and, instead, the meaning of events were turned into objects – reified – in the hope and expectation that the reified object was ontologically analogous to the objects of nature. Durkheim, who was among the first to propose that "social facts be treated as things", vehemently denies that this move affects the ontological status of the object under consideration:

> ...our principle implies no metaphysical conception, no speculation about the fundamental nature of being. What it demands is that the sociologist put himself in the same state of mind as the physicists, chemists, or physiologists, when they enquire into a hitherto unexplored region of the scientific domain. (Durkheim 1895, 58–60)

But, *pace* Durkheim, one cannot have it both ways. If a social fact is not a thing - an object that, according to Durkheim 'we know from without' as opposed to objects 'we know from within' – it cannot be known the same way as a thing. The ontological difference between objects of nature and objects of thought cannot be overcome by placing both on the same epistemological footing. If 'the state of mind' is not appropriate to the object of its contemplation, the result will be a misperception. The device with which this same state of mind is obtained is the ontological transformation of the object of study by renaming and reshaping the experience generating it:

> ... the sociologist, either when he decides upon the object of his reasearch or in his investigations, must resolutely deny himself the use of those concepts formed outside science and for needs entirely unscientific. He must free himself from those fallacious notions which hold sway over the mind of the ordinary person, shaking off, once and for all, the yoke of those empirical categories that long habit often makes tyrannical. (Durkheim 1895, 4)

Scientific status obtained by first admitting, as Durkheim does, that social facts are special, *sui generis* and differ from facts of nature, but next requiring that they be first transformed into 'scientific facts' in order to receive scientific analysis, is a status obtained under false pretenses. It is, in short, truth by stipulation. (3).

It is this illegitimate epistemological move that transforms the subjective experience of the second cup of tea into a 'fact', now the measurable and mathematically representable unit of marginal utility. The failure to achieve the level of predictive and explanatory power in relation to these fruits of reification, achieved by the natural sciences with respect to natural objects, is attributed to some sort of procedural error, to be overcome with the help of mathematics. If the marginal utility schedule could not find empirical confirmation, the problem was not with the transformation of the ontolgical subjectivity of the second cup of tea into a unit of measurement, but with the techniques employed in carrying out the transformation, Or so it was thought. But reliance on maths only compounded the problem by further constraining reification into the very narrow space where only those of its products could be admitted that could be mathematicised. Incommensurability was removed by a sort of intellectual fiat. 'Rational' and 'logical' was restricted to mean causally linked ends and means putting the notion of causality in the social sciences under a disfiguring strain. The 'utils' of utility theory said farewell to Aristotelian use value, and, with it, to the experience, including the moral meaning of usefulness. (Meikle 1995)

The XXth century obsession with the unity of science on the one hand and science being the only valid form of knowledge on the other has led the discipline to the construction of a surrogate reality with which a domain consisting of the sorts of valueless facts that comprise the natural sciences could be delineated. If, as John Stuart Mill (1893, Bk. VI, Ch. IV) famously put it, "Any facts are fitted, in themselves, to be a subject of science, which follow one another according to constant laws; although these laws may not have been discovered…" economics could achieve the much desired scientific status if it could define a domain consisting of such facts. Bur are there such facts, or would they have to be crafted, and if they have to be crafted what means could be employed to this end? Are they found, or are they contrived so as to meet the specifications of a propounded generalisation? In the latter case they amount to a surrogate reality, and, as some of the essays in this book show, that surrogate reality is an ontological falsehood. In a last ditch effort to save what could be saved of the myth of the unity of the natural and the social sciences. Emile Durkheim famously conceded that the social disciplines have their own, distinct subject matter, but, despite the difference, the same methodology applies to both. He follows Mill, who held that the explanation of natural and social phenomena had to have the same logical structure. The next step in this process, namely that the techniques of isolation, abstraction and idealisation could be freely imported from the natural sciences was a logical, if mistaken step, taken as if the absence of consciousness in a water molecule could be replicated by excising it from social facts. Peter Winch (1958), in rejecting this view, as indeed several of the authors in this volume, sees the mistake as one of failing to see that social explanations require the application of different concepts:

> …. the conceptions according to which we normally think of social events are logically incompatible with the concepts belonging to scientific explanation….the former conceptions enter into social life itself and not merely to the observer's description of it. (Winch 1958, 95)

Several authors in this volume, defend a broader and deeper claim: methodology must be appropriate to the ontology of the subject to which it is applied, and, given that the subject matter of the social sciences is ontologically different from that of the natural sciences, its methodology should also be different. The issue, therefore is not the *logical* incompatibility of the concepts being used, but, rather, the ontological difference of the subject matter between the objects studied by the natural sciences and the events examined by the social scientist.

It is difficult to overstate the consequences of this expulsion of intentionality from the reality with which social science is concerned. Having been removed from the reality under examination pursuant to the Durkheimian directive, the amputated 'fact', far from acquiring the characterstics of a fact of nature, is now open to the legislative manipulation of the economist. Leon Walras openly acknowledges this objective:

> ...the pure theory of economics ought to take over from experience certain type concepts, like those of exchange, supply, demand, market, capital, income, productive services, products. From these real-type concepts the pure science of economics should then abstract and define ideal-type concepts in terms of which it carries on its reasoning.*The return to reality should not take place until the science is completed and then only with a view to practical applications. Thus in an ideal market we have ideal prices which stand in exact relation to an ideal supply and demand.* (Italics added) (Walras 1926, 71)

Economics then is not concerned with understanding an existing reality, but rather, with the construction of a surrogate - in Walras' words, 'ideal-type' - world to which known reality should be subjugated. Markets should be regulated not in accordance with the intentionality of human beings - their will, desires and hopes - but in compliance with the ideal market of the pure science of economics. (4).

Surrogate realities may, of course, have great value. Like fables, parables and allegories, they can reveal important truths, and, in fact, modern economics does throw light on the complexities of economic life. They do not, however, contain scientific, law-like, universal regularities. Like fables, parables and allegories, the ideal-types of Walrasian economics arrive at their truths by contriving their subject matter. The difficulty with modern economics is not its enchantment with model building but with its conceit that models reveal the sorts of laws that the methodology of the natural sciences reveals about physical nature. The task, therefore, is a restatement of the aims and the domain of the social sciences in general and of economics in particular.

What should be the proper subject of economics? How are we to embrace the intentionality of economic life? Analogising social reality to physical reality, searching for questions and answers that rhyme with those formulated in the natural sciences is a category mistake. There is no nature in economics and no economics in nature. The proper subject of economics, therefore, cannot be the search for nature based law-like generalities with reliable predictive and explanatory power, because there are no such generalities to be had; instead, it should be the study of how human beings make decisions about the material needs and requirements of their existence. In delimiting the proper domain of economics, the following should be taken into account:

(i) There are no brute facts; economic events are doubly value-laden, first because the event under consideration comes about as a consequence of value judgment and second, because the event triggers a value-laden response;

(ii) There are no causal relationships - in the sense in which the term is meant in the natural sciences – among economic events. These relationships come about by virtue of human agency. The agency function may be substituted, as a short-cut, by a concept of abstract causation in models and in discourse about the economy to reveal a frequently recurring agency role, but it is a potentially disastrous mistake to take the one for the other. Turning agency into causation is the road to the elimination of the moral autonomy of the agent, and that elimination is the road to totalitarianism;

(iii) The economy and human agency within it is regulated not by laws, but by *rules and the roles formed according to* the paradigm governing the social unit of which the economy in question is a part. These rules may be quite open ended or they may be rigidly determined, or anything between these extremes. These rules and roles are inherently reflexive, constantly undergo change, and their discovery is the joint task of ethics, sociology, anthropology, jurisprudence and economics.

The subject matter of economics then is not the formulation of law-like generalities with which technocratic predictions can be underpinned, but, rather, the *interpretation* of the roles and rules giving rise to economic action. It may well be, as MacIntyre (2007) has it, that the human being is "a story telling animal" who directs his actions teleologically, that is to say, towards a goal, and does so with the aid of a narrative, or, in the alternative, Charles Taylor's (1985) emotional mode of perception as the foundation of human action, but from the perspective of interpreting economic reality, these premises are too distant to be useful. Instead it is proposed, that the acts of the agents should be understood in the context of the agent's relevant authorised role. What matters in understanding the purchases of an individual, or the decisions of a corporate official is their understanding of their roles as members of a household of certain characteristics or the duties and authority, the expectations, rewards and punishments associated with that corporate role. The overwhemling importance of roles and rules is particularly significant in the case of the modern corporation, the dominant economic agent of our times, created as it is entirely by legislation and governed by internally and externally imposed rules. The economy is much better understood through the rules legislated for any and all economic activity than with the aid of any theory about human nature even if these rules do not wholly determine economic activity, even if human judgment retains an ever active role, and even if, most importantly, those rules and roles contain and reflect very important socially and individually constructed value judgments, mandating at times the compusory excercise of such judgments. Rules and roles do not necessarily determine the outcome of economic action, – nor should they – but they do delimit its domain. Instead of determinist predictions, the search should be for propensities and tendencies formed in the light of the purposes human beings intend to achieve through economic means.

The proper domain of the discipline of economics then is the study of the ways and means with which individuals and groups decide what they should do to secure the material conditions of a *worthwhile life*. The requirement that the application of the theory must result in bringing about a worthwhile life rather than the maximisation or optimisation of any single issue also means that its tenets should be open to critical assessment on the basis of incommensurable considerations. Theories that lead to the degradation of human dignity or the environment will not do even if they satisfy the criteria of some algorithm. They are inadaquate not only because they are morally objectionable, but, more broadly, because they are ontologically deficient, that is to say, they fail to set out an adaquate picture of human reality.

14.1 Notes

(1) The inextricablity of values from facts is set out in Book I of the Old Testament. Eating the apple, now the ubiquitous trademark of our principal tool of knowledge, endowed Adam and Eve with the knowledge of good and evil. For the aetheist Hume the valueless fact may have seemed self-evident, but knowledge without the knowledge of good and evil is a modern idea. Contrary to the view that seeing all knowledge as pervaded by values is a post modern sceptical notion, the claim advanced here is that the inseparability of facts and values is shown in Genesis 2:17 to be the foundation of human ontology. It, eating the apple, is the defining event of becoming human.

(2) Facts recognised by Charles Taylor as value neutral are seen as synthetic or contrived facts hereinabove, synthesised so as to exclude values. They are surrogates, contrived to perform a desired function. For Charles Taylor's view of the facticity and value neutrality of statistical data, see Taylor, C. (2007) *A Secular Age.* Harvard University Press.

(3) This, the dominant form of reasoning in early XXth century French social thought is reflected in Walras "...the *pure theory of economics... is a physico-mathematical science...*the physico-mathematical sciences, like the mathematical sciences.. do go beyond experience as soon as they have drawn their type concepts from it. From real-type concepts, these sciences abstract ideal-type concepts which they define, and then on the basis of these definitions they construct *a priori* the whole framework of their theories and proofs."." Walras, L. (1926) (English translation 1954) *Elements of Pure Economics.* George Allen & Unwin, p. 71.

(4) For the connection between this ideology and the birth of XXth century totalitarian thought see Arendt, H. (1951) *The Origins of Totalitarianism.* Schocken Books.

References

Berlin, I. 1996. *The Sense of Reality*. New York: Farrar, Straus and Giroux.
Durkheim, É. 1895. Les regles de la méthode sociologique. In *Emile Durkheim Selected Writings, 1972*, ed. A. Giddens. Cambridge: Cambridge University Press.
Israel, G. 1996. *La Mathématisation du Reel*. Paris: Editions du Seuil.
Kornai, J. 1971. *Anti-Equilibrium*. Fairfield: Augustus M. Kelley, Publishers (1991 reprint).
MacIntyre, A. 2007. *After Virtue*. University of Notre Dame Press.
Meikle, S. 1995. *Aristotle's economic thought*. Oxford: Oxford University Press.

Mill, J. S. 1893. *A system of logic* (Book IV, Chapter 3).
Taylor, Ch. 1985. Self-interpreting animals. In *Human agency and language, philosophical papers*, vol. 1. Cambridge: Cambridge University Press.
Winch, P. 1958. *The idea of a social science and its relation to philosophy*. London, Routledge: Keegan Paul, Ltd, Humanities Press.

Index

© Springer International Publishing AG, part of Springer Nature 2018

P. Róna, L. Zsolnai (eds.), *Economic Objects and the Objects of Economics*, Virtues and Economics 3, https://doi.org/10.1007/978-3-319-94529-3

The manufacturer's authorised representative in the EU is Springer Nature Customer Service Centre GmbH, Europaplatz 3, 69115 Heidelberg, Germany. If you have any concerns regarding our products, please contact ProductSafety@springernature.com

Printed and bound by CPI Group (UK) Ltd, Croydon, CR0 4YY
06/10/2026
02229006-0002